The Masked Bobwhite Rides Again

John *Alcock*

The Masked Bobwhite Rides Again

THE UNIVERSITY OF ARIZONA PRESS

Tucson and London

Illustrations by Michael McCurdy

The University of Arizona Press
Copyright © 1993
Arizona Board of Regents
All Rights Reserved

⊛ This book is printed on acid-free, archival-quality paper.
Manufactured in the United States of America.

98 97 96 95 94 93   6 5 4 3 2 1

Library of Congress Cataloging-in-Publication Data
Alcock, John, 1942–
    The masked bobwhite rides again / John Alcock.
        p. cm.
    Includes bibliographical references and index.
    ISBN 0-8165-1387-2 (cloth : acid-free). —
    ISBN 0-8165-1405-4 (pbk. : acid-free)
        1. Natural history—Sonoran Desert.   2. Sonoran Desert.
3. Desert ecology—Sonoran Desert.   4. Cattle—Sonoran Desert—
Ecology.   5. Man—Influence on nature—Sonoran Desert.   I. Title.
QH104.5.S58A39   1993   93-15416
508.791'7—dc20   CIP

British Library Cataloguing-in-Publication Data
A catalogue record for this book is available from the British Library.

Contents

Preface   vii

DESERT MOUNTAINS

A natural history   3

Cactus-hugging in practice   7

Army ants   10

Death comes for the black-tailed gnatcatcher   13

Michael's ashes   17

The birth of a paloverde   21

DESERT PEOPLE

Schilling's best   33

The miner's cat   36

Where did all the glyptodonts go?   41

Thirty-eight Apaches   51

The last Indian war?   61

Bandidos   69

Confessions of a cactus-hugger   76

Abert's towhees and other opportunists   82

Playing God with the white-winged dove   89

DESERT CATTLE

Cows   97

One man's campaign   102

Cattle free in 1893   107

The impact of an impact statement   116

Mountain lion mathematics: A report from
    Klondyke, Arizona   121

More mountain lion mathematics   127

Cowpies   128

Peccaries   133

Death in a saguaro forest   137

Randolph Canyon and Burro Creek   141

The masked bobwhite rides again   147

DESERT HOPE

Life in a saguaro forest   157

The black bear in Ballantine Canyon   160

The Mazatzal Wilderness Area   162

The coyote in South Mountain Park   164

December rain   172

References   175

Acknowledgments   181

Index   183

Preface

> The world today is sick to its thin blood for lack of elemental
> things, for fire before the hands, for water welling from the earth,
> for air, for the dear earth itself underfoot.
>
> HENRY BESTON, *The Outermost House*

In the first part of this century, Henry Beston was able to find
in the dunes of Cape Cod, Massachusetts, the elemental things so
necessary to maintain the human spirit. Today he would have a
hard time finding a parking place on Cape Cod, let alone the solitude required to come to grips with the dear earth. But I think
that if he were with us today, he would find reason for hope in the
desert regions of the arid West. There are places here where it is
still possible to entertain the notion of aloneness, wildness, natu-
ralness, where it is still possible to imagine a landscape not totally
dominated by our fellow man.

As a born-again Arizonan transplanted in stages from Massachusetts to the Sonoran Desert, I have adopted my latest home with
all the enthusiasm of a religious convert. Each time I venture into
the desert, I can still barely believe my good fortune in being able
to visit places where you can walk without bumping into other
people, where the earth is populated with so many strangely wonderful plants and animals, each one competing for the honor of the
most beautifully adapted to a most rigorous environment or most
unEastern in its aspect or most ascetically aesthetic.

In this book I have tried to convey what it is about the natural
history of my adopted homeland that appeals to me so strongly.
I shall also explain why the flood of human immigrants that have

joined me in the desert and the cows that call this land home, too, cause me great concern and worry. There is, I believe, ample reason to fear for the integrity of the Sonoran Desert in central Arizona, to wonder if we will maintain the biological heritage that preceded us here and has the capacity to enrich the lives of us all, if we would just respect it a little more.

I express both my satisfactions and concerns as one person speaking for myself and not representing any group or institution, certainly not the one that employs me (the Department of Zoology at Arizona State University). In the course of the book I write from my own experiences and offer my own conclusions, which are not flattering when it comes to Arizonan ranchers or the cows they run on Arizona's federally administered public lands. I realize that some persons may take offense, most notably that element of the ranching community that possesses grazing leases on public lands.

For what it is worth, let me say that I am aware of the controversial nature of certain of the arguments I present here, and I have tried to flag these statements in the text so that the reader will be aware of their controversial element, too. Moreover, I have no doubt that many Arizona ranchers have great affection for the desert and that some are as ecologically aware as the bovine-bashing environmentalists with whom I feel a greater allegiance. My goal is not to stereotype ranchers as the bad guys in this drama but to question whether the current state of affairs is really the best we can do in fulfilling our responsibility to respect and maintain a truly remarkable place.

I have some optimism that we will improve our record in this regard. There are some success stories to report, and I devote the final section of my book to them. The desert has a certain resilience as well as the depressing fragility for which it is better known. Let us permit desert ecosystems to demonstrate their resilience, their capacity to come back. Let us acknowledge the value of the desert for things other than unadulterated commerce and extractive enterprise. Let us exercise the power of restraint. In so doing, we might avoid the fate that Beston feared would befall us, that of becoming cosmic outlaws, "having neither the completeness and integrity of the animal nor the birthright of a true humanity." It is a birthright worth preserving.

The Masked Bobwhite Rides Again

DESERT MOUNTAINS

There is a pleasure in the pathless woods,
There is a rapture on the lonely shore,
There is society, where none intrudes
By the deep Sea, and Music in its roar:
I love not man the less but Nature more
LORD BYRON, Childe Harold's Pilgrimage

A natural history

The dry wash that drains the gentle canyon to the southwest of Usery Peak passes through a narrow, rocky throat before spreading out again on its static journey toward the distant Salt River. At the pinched entrance to the canyon, several strategically placed boulders act as an impassable barrier to off-road vehicles. The dead-end boulders regularly confront four-wheel-drive Toyotas, Chevy trucks jacked up so far that the front door handle is eight feet off the ground, and even motorcycles with the air let out of their tires, which zoom up the pebbly avenue swerving past the projecting limbs of ironwoods lining the wash. The off-roaders stop when they can go no farther and turn their machines around by the boulders. Most linger for a while, some to riddle assorted targets with ten dollar's worth of Winchester .22 longs, PMC sidewinders, or Remington thunderbolts, some to dispose of a twelve-pack of Coors, some to construct a fire ring in the middle of the wash, fueled with punky paloverde limbs hacked from a nearby tree. Sooner or later, they are on their way again, lurching down the wash in their ORVs toward reassuringly firm pavement.

Almost no one scrambles over the boulders to walk up the wash, surmounting the obstacles that continue to obstruct travel for a half mile or so. At least I do not find many footprints today after having hiked in by a circuitous route of my own. I do find a neatly carved groove, nearly three feet long, traveling down the face of one of the big gray stones that forms a dry waterfall in the middle of the streambed. It is hard to believe, but water has flowed down the wash with sufficient force a sufficient number of times to gouge out a silky smooth spout in solid rock.

Overhead a great horned owl flaps slowly from one ledge to another on wings as broad as they are long. Back-lighted against the morning sky, its calm, silent flight seems part of the undisturbed atmosphere of this delightfully uncivilized place.

Eventually, the drop-offs and rock walls give way to more open terrain, completely inaccessible to off-road vehicles. Here the wash has formed a broader, smoother trail that ascends at a comfortable angle. A single dried desert mint, now reduced to a skeletal memory of the past spring, still stands upright on a sandy shelf in the middle of the waterless stream.

Teddy-bear cacti gather in groups on the shoulders of the wash. A house finch settles gently on one of the cacti ignoring multitudes of spines that seem to offer a hostile landing spot at best. Here and there, towering well over the three- and four-foot teddy-bears, stand the Sonoran Desert's most famous plant, saguaros ten, fifteen, and twenty feet tall. A few of these immense cacti grow right in the wash itself, having reached a size capable of withstanding the thunderstorm thrown surges that come down the streambed at great and irregular intervals. The saguaros hold their arms up in postcard poses.

Some fifty feet from the wash, a monster saguaro that has lost its picture-perfect innocence occupies a small depression in the rocky soil. This cactus's immense arms have drooped to touch the ground before turning up again. A severe frost long ago weakened the tissue in the saguaro's arms, causing them to bend but not break, and in the years since the disaster, the plant's customary response to gravity has induced an upward turn in the still living but fallen appendages.

Somewhere along the line, perhaps due to freeze damage, the top half of the central trunk of the saguaro fell to the ground and decayed so long ago that no sign of its existence remains on the gravel beneath the decapitated cactus. Despite its mutilation, the saguaro seems reasonably healthy except for its flower buds, which have attracted consumers, perhaps small pocket mice, whose burrows riddle the ground around the cactus. In any case, something small and agile has managed to clamber onto the arm tips and gnaw into the buds. The damaged specimens soon turn black before falling from their perches onto the little bursages squashed between the saguaro's elbows.

The mammoth saguaro, through the accidents that removed its upper trunk and lowered its arms, will have no surviving fruits at all this year. Without fruits, it will produce no seeds. Without seeds, it will produce no new offspring. But even if it misses the annual opportunity to leave a descendant or two, it seems to be in no danger of dying. It looks capable of living on forever, occupying its special patch of desert, gleaning water from the soil after a rare storm, propping up the dead ironwood that once grew near it.

There is a timeless aspect to desert landscapes populated with huge saguaros, leaving the viewer with a sensation of an immortal and unchanging world. Which just goes to show that sensations can be highly misleading. First, although saguaros are without dispute capable of living a long life, most do not survive for a great deal longer than the average American citizen. To hang on for 200 years, as the mammoth cactus may have done, requires an exceptional run of luck and the capacity to tolerate a great many indignities.

Moreover, although it was once thought that the Sonoran Desert's vegetation had occupied its modern geographic range for perhaps 20 million years, in reality there have been big changes here relatively recently in geological time. Had you or I been fortunate enough to visit the western flank of Usery Peak 13,000 years ago, we probably wouldn't have seen a single saguaro cactus. I can make this assertion thanks to long-deceased woodrats of various species whose ancient middens provide clues about the past, clues that were unknown until just a few decades ago.

Woodrats are mid-sized rodents that have occupied the western United States for thousands of years during which time they have created durable mementos of their lives in the form of piles of trash. These middens, as they are called, are the result of the excretory habits of woodrats, which use a special cranny or recess in their dry cave homes as a bathroom. The bathroom is also a trash heap that receives half-eaten plant fragments and other inedibles in addition to the animal's feces and urine. The combined debris eventually forms a lacquered, urine-impregnated package, which can become a fossilized testament to the meals and excretory capacity of woodrats that lived in a long-gone millennium.

Some early pioneers on their hard journey through the arid west hoped they had found manna when they came upon woodrat middens, and they actually ate some midden fragments, having been deceived by their supposedly candylike appearance and texture. No modern Westerner who has seen a woodrat midden can easily imagine how one could take the first bite, let alone persist in eating a midden candy bar. But then again, few modern Westerners have been as hungry as the midden munchers doubtless were.

The use of middens as food was a short-lived phenomenon, but in 1964 Phil Wells and Clive Jorgensen realized that these materials might serve another function for us all, namely as samples of vegetative life during the years when the middens were formed. Radiocarbon datings can be readily secured from the organic material in a midden. Once the plant fragments and fossil pollen that the rat collected 1000 or 10,000 or 20,000 years earlier have been separated, identified, and dated, the result is a snapshot of the plant life from a world that no longer exists.

Arizona middens tell us that the Sonoran Desert some 11,000 to 15,000 years ago was dramatically different from the current desert world. For example, middens made more than 10,000 years in the past lack saguaro seeds and brittlebush remains and instead yield fragments of pinyon and juniper trees, Mojave sage and Joshua trees, as well as numerous grasses, all plants that now occur far away at elevations hundreds of meters higher. The landscape of this time, instead of being covered with Sonoran desertscrub, was reminiscent of habitats now found at elevations several thousand feet higher, where scattered junipers and pinyon pines grow on grassy slopes and meadows.

As recently as 11,000 years ago, southern Arizona was evidently cooler and wetter than it is currently, creating a climate suitable for plants that today survive only at higher elevations. Saguaros did not arrive until 10,500 to 9,000 years ago. They are believed to have come up from an arid refugium in Mexico, as the weather turned slowly drier and warmer throughout the Southwest.

The transformation of the region took place gradually. The first saguaros to reach Usery Peak probably coexisted with junipers, and perhaps shrub live oaks, for hundreds or even thousands of years. The continued drift toward extreme aridity and higher temperatures resulted in the eventual disappearance of these trees from lower elevations in the Sonoran Desert and their replacement with paloverdes and ironwoods. But it was not until around 4000 years ago that the Sonoran Desert of Arizona adopted its current aspect in which paloverdes, bursage, creosote, and a host of cacti, among them saguaros, dominate the land. Despite appearances to the contrary, the current desert environment is young in geological terms, barely formed, yet formed well enough to obscure a past that might have been lost forever were it not for the durable labors of desert rats.

Looking out across a vast bajada cloaked in eye-catching saguaros and a supporting cast of hundreds of paloverde trees, my fellow cactus-huggers and I might conclude that this is the way it has always been (and always should be). We are entitled to our opinion on how it always should be, but if the past is guide to the future, the woodrats occupying the Userys several thousand years from now may well be urinating on midden heaps laced with bits and fragments of novel plants that do not occur here today. Perhaps there will also be a few plant-huggers around to admire the new species that have replaced the diverse and wonderful vegetation of the present. But the persistence of humans is far from assured, and so we, the living, had better enjoy the present desert landscape while we can.

Cactus-hugging in practice

Taking my own advice, I set out to enjoy the modern Userys once again. Named for a rancher who once occupied the area, the Userys are not tall, impressive, or exceptionally rugged as Arizona mountains go, but I love them anyway. The rains that very occasionally fall on these hills have slowly cut gullies and little canyons in the mountainside. From the gullies and canyons come thin tongues of sandy gravel, which lick their way across the desert flats. I regularly use one or another of these washes as an access route into the mountains.

Each time I hike up into the Userys and back down again, I know that I am walking through a land that has been through a host of changes, subtle and not-so-subtle, natural and not-so-natural, some caused by slow-moving but inexorable geological and biological processes and others by fast-moving humans and our heavy-footed livestock. To appreciate the desert fully, we probably have to come to grips with all these changes, a task that requires some accommodation on the part of cactus-huggers.

Today's walk into the Userys starts on the other side of the mountains far from the wash trafficked by off-road enthusiasts. Here the meandering wash that does double duty as a trail angles through a nice stand of saguaros, perhaps one whose members are descended from some pioneering cactus that set its roots in the desert soil here 6000 or 8000 years ago. A coyote slips away among the cacti and paloverdes, stopping to take one last look over his pale brown shoulder before evaporating in the desert heat.

The trail leaves the wash and climbs steadily higher with intermittent drops into and then out of eroded gullies that have provided the fine gravel which carpets the washes below. A final hard uphill pull and I am on the top surveying a great chunk of central Arizona. Mountains far more imposing than the Userys define the northern and eastern horizons, while isolated mountain islands poke up from the southern and western plains.

Much closer to my perch high on my own mountain island, a downhill saguaro with a great looping arm catches my eye. This is one I have never visited, and I set off down the mountainside, picking my way cautiously over the loose rocks and among the jumbles of boulders that lie between me and the cactus, keeping my eye on the teddy-bear chollas, which present an intimidating coat of white harpooned spines.

The slow descent is uneventful until I begin to cross a slope covered with walnut-sized gravel. Shortly thereafter my trip accelerates dramatically as I tumble down, thumping into the stony earth. I am surprised at how rapidly I make the transition from an upright to a recumbent position. Reaching blindly down to brake my fall, I slam my right hand directly onto a teddy-bear cholla joint lying amidst the gravel. It is one of many fat bratwurst-sized chunks of cactus that have fallen from their parents and now lie haphazardly on the desert floor. A few have taken root and begun to form a new generation of this most spiny of cacti.

The joint that has attached itself to my hand appears to have taken root there, so firmly are its long straight spines embedded in my palm. I peer at the thing, half in shock, half in disbelief. Looking about I find a thick stick with which I slowly pry the elliptical joint out of my flesh. It leaves behind roughly fifty broken cactus spines. I spend some time plucking these one by one from the palm of my hand. A number refuse to come free completely, breaking off to leave the tip in my hand. These will have to wait for later extraction when I have access to tweezers and needle.

Eventually, my palm, although throbbing and peppered with red dots, is relatively cactus free. Sitting on the ground, I only now notice that my trousers are ripped and my knee cut as a result of my fall. I reflect on the vicissitudes of life for awhile, trying to recover enthusiasm for cross-country hiking before proceeding.

I cover the short remaining distance to the unusual saguaro without incident. Two white, waxy flowers perch right at eye level on the end of the huge downward looping arm. A small black bee lands and slips into one of the flowers. A nearby saguaro completed flowering long ago, and now its arm tips are covered with an array of large red fruits. A trio of house finches perches by the opened fruits to remove beakfuls of the crimson interior flesh and tiny red-purple seeds.

Leaving the distinctive saguaro behind, I reach a sandy wash that drains a substantial sector of the mountain. It wanders downward at a more comfortable angle than the surrounding hillside, spreading out in a delta of fine gravel in places, constricting in others where rock walls funnel the dry watercourse into narrow chutes.

At the edge of the wash, a group of five finches, one ash-throated flycatcher, and a pair of black-throated gnatcatchers have assembled at an untidy, leafless wolfberry shrub whose black limbs create a thicket above the sand. The birds seem to be peering down into the shrub. Even as I come close to the spot, the gnatcatchers stay by the wolfberry until finally fluttering away with wheezy cries of alarm. A diamondback rattler slides fluidly beneath a tangle of limbs to coil itself neatly in a sandy depression within the shade of the plant. Its tail gives a little shiver of sound, then falls silent.

In the gentle canyon formed by the wash, most of the saguaros have begun to produce ripe fruit. The fruits stand side by side on arm tips and trunk tops, some opening neatly to expose a rude black tongue of clustered seeds while others have split raggedly, their contents already torn and partly eaten. A strong gust of wind sweeps down the valley, rocking the saguaros; at the same time a wild musical wailing fills the air. For a fraction of a second the cries seem to be the music of the wind whistling through the arms of the cacti. But the wailing builds in volume and complexity even as the pulse of wind disappears to parts unknown. The wild sounds come from a chorus of coyotes, which has gathered for the occasion only a few hundred feet away, judging from the intensity of the yelping. For half a minute the coyotes yip, howl, wail and yodel in their melodious and exhilarating fashion. Then the concert stops in mid-note. In the abrupt silence that follows, the saguaros, palo-verdes, and spiny chollas seem frozen in place, as if they were listening for the coyotes to complete their unfinished but nearly perfect desert song.

Army ants

A coyote concert is in my experience a rare and treasured event, but it is only one of the many possible desert moments that make walking up into the mountains worth the effort. It is true, however, that there are usually some obstacles to overcome on such a walk. For example, the faint trail along the northern exten-sion of Usery Mountain begins by wandering up a rise littered with plastic milk bottles and cardboard boxes shredded recently by gun-ners and left in place as testimony to their pastime. Scallops of pale gray plastic lie among the bullet-scarred rocks. A paloverde in the line of fire sports amputated limbs with pale yellow stumps. The green skin of its still intact limbs is peppered and pockmarked with ricochets and fragments of bullet-blasted gravel.

Although the low-elevation paloverde is decidedly the worse for wear, most other paloverdes higher on the mountain have had the good fortune to avoid the attentions of Homo ballisticus. The peren-nial plants as a group look much the same as they did a year or two ago, except perhaps for showing the effects of this year's re-duced winter rainfall. In this spring of drought, the brittlebush by the trail have one flower stalk for every ten that they produced in wet springs past. Now the season of flowering is over, and the few stalks of the year have turned brown while the petals of the flowers they supported have long since disappeared. Next to the somber brittlebush at trail's edge, a bright red shotgun shell looks even more out of place than usual.

Higher still, the path along the backbone of the ridge is cut pre-cisely at right angles by a column of ants streaming from left to right. Although the reddish ants are individually small, they run three, four, five abreast creating a wavering line a half-inch or so thick. The ants carry in their jaws bits and pieces of some other insect, adding extra mass to thicken the body of the column. Tracing the line of ants backwards leads me soon to a colony of harvester ants that has occupied this patch of desert for many years now. The harvesters maintain a large cleared apron around their nest, keeping this area mowed and weed free like an obsessed suburbanite. Today bodies of the once house-proud harvesters dot the pale white pebbles by the nest, creating a mosaic of death and dismemberment. The little army ants, for that is what they are, have apparently completely de-stroyed the harvester colony in a massive assault overnight. Now in the aftermath of what must have been a titanic battle, they are in the midst of cutting up their victims, subduing any residual re-sisters, and hauling off their prizes. In organized chaos, little soldier army ants drag thoraxes and abdomens of their much larger victims out of the harvester ant burrow and march off with them briskly, leaving a swarming mass of their fellows at the nest entrance en-gaged in cleanup details.

Army ants of this species (*Neviamyrmex nigrescens*) march out time and again from a temporary underground bivouac that remains in place for several days. During this time, the queen stays behind in her bunker while her sterile daughters, the worker-soldiers, forage for victims elsewhere. As the several columns of soldiers weave loosely across the terrain, one or another line may run into a nest of fellow ants, upon which the deadly raiders converge. After de-pleting the colonies of prey in one area, the army ants pack up and move to a new location to repeat the cycle all over again. The point of their overwhelming assaults is to gather food to feed themselves and the pale larvae produced by their queen mother.

The raid that I observe will have repercussions for this small patch of desert ridgeline for years to come, in years of drought and years of plenty. The death of this one harvester ant colony cre-ates vacant real estate that small annuals and other plants may soon occupy, now that the fastidious harvester ants no longer exist to weed their nest apron. The seeds of plants found many meters from the nest will not be gathered by these energetic harvesters, since they themselves have been harvested by their fellow ants. Pocket mice and black-throated sparrows will be among the beneficiaries.

The gains of mice and sparrows may well be only temporary be-cause eventually from other still extant colonies of harvester ants, mated queens will come pioneering after their late summer nup-tial flights. Those that happen to dig their initial nests into this bit of ridge will not encounter the competition for food that the huge worker force, present in the old established colony, would have provided. Most of the pioneers will die anyway, the new queens unable to secure sufficient seeds to survive as they try to produce a generation of daughters, which will become the first worker helpers in these infant colonies.

Perhaps a foundress queen and some of her brood will succeed in keeping the colony alive for a year or two at this place, the harvest of the workers providing energy for the creation of a still larger work force dedicated to gathering ever more seeds, defending the nest, and rearing new batches of their mother's brood. Then, when walking this part of the ridge you will have to watch once again where you put your feet so as not to acquire an overly officious worker eager to implant its formidable stinger in human flesh. A harvester colony will be back in place, vigorous, seemingly invulnerable, apparently immortal—were it not for events like an army ant raid or the queen's inevitable death, which will keep the community of ants and seed-producing plants constantly dancing up and down this ordinary desert ridge.

Death comes for the black-tailed gnatcatcher

The demise of a single colony of harvester ants can have a kind of ripple effect on a host of other members of the desert community. The same is probably true for a great many other desert animals, with death providing the impetus for change far beyond the elimination of a single living entity.

But it is hard to discern much of cosmic importance in the body of the moribund black-tailed gnatcatcher crumpled on the stony ground beneath a paloverde tree in the Userys. Gnatcatchers are small birds and death has compressed this bird even more. It lies on its side, eyes evaporated, feathers disheveled. When I pick it up, I find that it weighs almost nothing in its mummified, rather than decayed, state. Arizona's heat is a dry heat, as we remind ourselves rather too often, and it has done a fine job of preserving the gnatcatcher's corpse. There is no obvious sign of the cause of death. I suppose that the bird, which appears to have been a juvenile, was just one of many this dry spring that ran out of luck and food after the first few weeks of its life, a dangerous and difficult time for animals of all sorts in all environments. Without its eyes it does not look pathetic or even terribly real, merely small and feathery.

As I think about it, I realize that I have never before found the body of a dead gnatcatcher in the desert. Not a dead verdin either, or warbler of any sort, or woodpecker, or dove. In fact, I am hard-pressed to recall ever having found a dead bird before except on streets and highways and on ocean coasts where defunct gulls, terns, and cormorants commonly find a last resting place amidst the flotsam of high tide. I have found a complete set of feathers of some birds, a Steller's jay on one occasion, a mourning dove on another, that had been plucked by a predator (I suspect a Cooper's hawk), the feathers loosely arrayed on the ground in a kind of memorial wreath for the departed bird. But finding entire bodies of deceased land birds is not an everyday event. (I exclude the cases of small birds that crash into large glass windows and expire of a broken neck. These unnatural deaths often occur within view or hearing of the occupants of the house, and the bird's body falls onto the manicured grounds outside the window, where it is easy to find, stimulating partly deserved guilt on the part of the discoverer.)

Why is it that we so rarely come across a dead bird? In many places and many seasons, land birds that died would fall to the ground and be immediately concealed by vegetation. Furthermore, although dead, the body of the bird contains calories and nutrients considered attractive by many scavengers—a snuffling opossum or skunk, the neighborhood mongrel or coyote, if the neighborhood is lucky enough to have a resident coyote.

Even smaller creatures can be effective recyclers of dead birds. A guild of carrion beetles has evolved the capacity to locate recently deceased birds and mice, which they bury by excavating the soil underneath the corpse. The body sinks into the cavity and is completely concealed after just a few hours of work by these diligent undertakers. The beetles then convert the hidden corpse into a nestlike ball of flesh, near which the female lays her eggs. The grubs hatch and receive regurgitated carrion from the male and female, which often stay together to rear their brood cooperatively on the food item that they buried. By applying a chemical secretion to the carrion, the beetles eliminate bacterial decay and the odor that would reveal the location of their bounty to the many mammalian scavengers that might otherwise follow the scent to the grave and disinter the corpse.

I mention these mildly macabre details to make the point that not only are dead birds a food resource for some animals, but they are a highly desirable one, so much so that there may be competition for recycling rights—thus, the rapid disappearance of the remains of deceased birds in many places.

The larger point still is that a scarcity of conspicuous avian corpses is no indication of the frequency with which birds shuffle off this mortal coil. Unlike human populations, which now grow and grow with relentless abandon, the numbers of most common bird species remain reasonably stable, although only when considered over the long haul. From year to year, populations expand, contract, and expand once again in a crazy-quilt pattern that reflects annual changes in the resources available for local populations of birds.

Evidence on this issue comes from a great variety of sources, including the results of day-long censuses made in many parts of the United States around Christmastime under the auspices of the Audubon Society. These counts are published annually, and they provide a crude measure of the abundance of particular birds from year to year.

A goodly number of Christmas counts, for example, have been conducted at the same Gila River site in central Arizona. The spot provides fine black-tailed gnatcatcher habitat, and the bird has regularly appeared on the lists assembled by dedicated bird-watchers on the census day. In the period from 1984 to 1990, the number of black-tailed gnatcatchers seen has jumped and fallen, only to rebound again. In 1984, bird-watchers tallied 58 black-tailed gnatcatchers with subsequent counts going from 19 to 104 to 49 to 80 to 61 to 116. Part of the variation in the numbers recorded derives from differences among years in the number of observers and total hours spent in pursuit of birds to tally. In general, more observers have come on board for the counts as the years have passed. Additional causes for annual differences in gnatcatcher numbers may stem from the vagaries of weather conditions on the days selected for the censuses and from differences in the desire of the observers to jot down every last black-tailed gnatcatcher they came across while searching for something rare and exciting.

Even when all these extraneous factors have been considered, it is close to certain that gnatcatcher numbers have not remained absolutely identical in the censused area from Christmas count to Christmas count. Droughts are as serious an impediment for reproduction for insect-feeding gnatcatchers as they are for seed-producing brittlebush.

Whatever the causes for the year-to-year fluctuations in population, Audubon Christmas counts offer no reason to suspect that black-tailed gnatcatchers are in the midst of a population explosion. In the long run, hatchings in this species are more or less matched by deaths, which means that each year a great many small gnatcatchers find a final resting place beneath paloverde trees where most will be found, not by me but by scavengers glad to have an easily obtained meal, albeit a small one. These gnatcatchers will not be available for the next Christmas Bird Count, but sad though this may be, there is harmony of a sort in a population at equilibrium, in the stern symmetry of checks and balances, removal and replacement, death and birth.

Michael's ashes

On the south-facing slope of Usery Mountain on a late after-noon in mid-June, only the paloverdes still seem to be alive. They provide feathery patches of faded green on a sun-bleached land-scape. The exuberance of a springtime fuelled by exceptional rain-fall during an el niño year is just a memory now. It was a year when the Pacific Ocean currents changed their routes, the storms off the California coast became more numerous and wandered in-land, and the desert vegetation flourished in response. But now the once-abundant, well-watered annuals and grasses, greener than the paloverdes only a short time ago, have dried into a somber array of pale browns and paler yellows.

I pick my way slowly up the mountainside, balancing on the boulders, slipping on the steep gravelly sections, brushing past frag-ile, dormant brittlebushes. Dead grasses crumble beneath my step, the fragments falling to the heat-saturated ground to complete their decomposition under the sheer weight of the sun. It is a long and cautious climb, four or five steps up, a pause, five or six more steps, a pull on the canteen. The sun has plenty of punch left, but in its descent its angled light brings some definition back to the land and pulls the pleated ridges of the saguaros out of hiding.

The ascent produces the mountaintop at last and a view of Four Peaks, reemerging monumentally from the gray haze that has di-minished the horizon during the middle of the day. Between Usery Mountain and Four Peaks lies another broken range of moun-tains tied together by a long yellow band of sedimentary rock, a stone ribbon, frayed and tattered by the erosive effects of time and more time.

Peccaries have evidently come to this ridgeline lookout on many occasions in the past, perhaps to enjoy the view, certainly to defe-cate. The innocuous fecal calling cards they have left behind range in age from a few days to many months old. All stages of oxidation are represented in the collection with colors to match, dried brown for the relatively fresh deposits, burnt gray for those of intermediate age, and ash white for the oldest, which have been sun-cured to the white achieved by an exhausted charcoal briquette.

A gang of turkey vultures, which roost on a giant rock pile at the eastern end of the ridge, are taking their late afternoon promenade, sashaying out from the roosting site only to turn back after having inspected the entire ridgeline to the west. Two vultures hang above my head like red-tailed hawks, taking advantage of a welcome late day breeze that is deflected upward by the slope. Their pale green feet and blood red heads struggle to bring a little life to their otherwise funereal plumage. The long, black primary feathers of their wing tips ripple in the wind. The afternoon works its way toward evening; heat curls up from the boulders crouched along the ridgeline beneath the drifting vultures.

On my return trip down the mountain, a mourning dove leaps up from the ground at the last minute, bursting out of a patch of dried grasses and brittlebush with a clatter of wings that sets my heart to jumping for a moment and tightens the muscles in my arms. Another dove, a white-wing, comes quietly down the mountainside, a gray streak uninterrupted by wing beats, obeying but controlling gravity's pull, curling over on its side to flash around a great pile of tan rock slabs, stacked in an orderly array in imitation of a megalithic tomb for a nameless Neolithic headman.

Shadows slip slowly down the long western incline to the valley. The dark patches grow imperceptibly, inching their way down to the wash, pooling in depressions, spilling around the paloverdes, slowly drowning the stubborn sunlit earth.

Near where the descending wash begins to dive among encroaching canyon walls, a huge saguaro lies flat on the ground. Michael's ashes, gray and white, rim the now exposed root ball. The saguaro, although dead for nearly two months, is barely deflated. The thin tips of its outstretched arms have begun to decompose but the rest of the cactus remains faithful to a remembrance of its living self.

About seven weeks earlier, five of us had driven to the dead end of the dirt track extension of Hawes' Road and walked by the shortest route possible into the hidden valley. Michael's wife, Kit, carried Michael's ashes, which had journeyed to Arizona in a sturdy, no-nonsense white box that had provoked curiosity among the airport x-ray operators. We hiked up a long ridge and down the other side to the giant saguaro, which was alive at the time, its outstretched arms raised optimistically in the air rather than lying in a jumbled mound on the ground. At the base of the standing cactus, Kit carefully poured out Michael's ashes, of which there was a surprising quantity, and smoothed them in among the bursages. My wife and I watched. Michael's son and Kit's son sat on the ground. A raven drifted morosely overhead.

The ceremonial morning was warm and pleasant. The brittle-bush, beneficiaries of the abundant rains of the winter and early spring, were extravagantly yellow, the round bushes with their crowns of bright flowers mushrooming up among the boulders, aligned shoulder to shoulder in the open spaces, crowding in among the prickly teddy-bear chollas, filling the gullies on the hillsides, gullies that still contained trickles of water.

Kit read from Michael's Arizona journal, having selected a passage that expressed his affection for the desert. The rest of us listened, the teenaged boys' faces impassive, their thoughts their own.

A rock wren bobbed and trilled from a distant perch.

Kit read a paragraph from Ann Zwinger's *The Mysterious Lands*: "I prefer the absences and the big empties, where the wind ricochets from sand grain to mountain. . . . I prefer the raw edges and the unfinished hems of the desert landscape. Desert is where I want to be when there are no more questions to ask."

The saguaro that we selected to be Michael's marker leaned to one side but otherwise was a wonderfully large and symmetrical specimen, rich in character and beauty. The five of us sat by it for awhile after Kit had finished speaking, our minds searching among the various conclusions, all of them incomplete, uncertain and unsatisfactory. Then the five of us retraced our steps up a steep hillside, stepping around the lupines and desert poppies and a dozen other wildflowers still in bloom. The boys and I lifted up some likely rocks and in so doing, exposed two scorpions, one large and exquisitely chunky, the other small and yellow, and a thin snake that seemed confused to see the morning sun. We replaced the rocks, covering up the scorpions and snake, making sure to recreate their shelters rather than flattening the uncovered fauna by mistake.

A couple of weeks later I returned to the Userys again but climbed the north slope to the ridgeline rather than wandering in the enclosed southern valley far below. With my back to Four Peaks, I looked down toward where I knew Michael's ashes were, scanning with my binoculars to see what I could see. What I saw was Michael's saguaro lying flat, arms thrown forward, dead, the probable victim of el niño with its surfeit of water, which the cactus had absorbed and stored, taking on excessive cargo at the expense of becoming fatally unbalanced.

My discomfort at the sight of the collapsed saguaro mingled with irritation and uneasiness. I had known the cactus longer than I had known Michael. Its death seemed to compound Michael's death, which was the end product of a statistically improbable but extremely real brain stem tumor. Medical attempts to combat the tumor succeeded primarily in stimulating valley fever, a fungal disease endemic to the Southwest. The dormant spores of the fungus had traveled back to Indiana in Michael's lungs after his sabbatical here; they had surged into activity when Michael's immune system had been compromised by the efforts to irradiate the tumor into oblivion. Valley fever greatly complicated Michael's treatment and helped make his terminal year ever less hopeful and ever more painful. As the months passed, Michael gave up the idea of a trip back to Arizona in the summer; he postponed the thought of resuming his academic career in the fall until the next spring; as the next spring approached, he postponed it once again; he lost the use of one arm; he went on disability; he made plans to have his ashes returned to Arizona; he lost clarity of speech and the ability to walk unaided; he lost his life.

Today Michael's ashes are visible only to someone who knows what to look for. The remains of the saguaro, on the other hand, will be highly conspicuous for years to come. Some of the drought-resistant flesh of the saguaro will linger for a couple of years. The structural internal ribs of the cactus will last much longer still, especially the thick, woody cylinder that once provided support for the plant's elegant crown of arms. It will be decades before these durable ribs crumble into ashes, probably well after I have ceased to come to pay my respects to the saguaro and to Michael.

The sun drifts out of sight, permitting dusk to have its moment. In twilight I walk out of the valley among the now-subdued brittle-bush and in so doing I disturb a solitary peccary. The pig dashes wildly away with leaps and snorts, a picture of vitality for the time being even if its brown pelt is askew, its thin body so laterally compressed as to almost achieve a single dimension. The pig's flight takes it out of sight almost instantly. The sound of peccary feet on gravel persists for a few seconds longer, followed by silence, an absence, a big empty.

The birth of a paloverde

A black-tailed gnatcatcher might live three or four years, if blessed with inordinate good fortune. Michael nearly accumulated a half-century before inordinate misfortune ended his life. Paloverde #17 has prospects for a lifespan of more than 100 years, maybe even 400, if the great desert botanist Forrest Shreve was right in estimating a four-century existence as a maximum for paloverdes.

Paloverde #17 stands slightly off center near the top of a minor bump on one of the many ridges that straggle out from Usery Peak. The little tree is similar to the one that served as a funeral bower for the dead gnatcatcher. Like most other mature foothill paloverdes growing in the Userys, it is a scruffy, unkempt plant, with a tangle of green branches and twigs radiating outward from a crooked central trunk. For much of the year the paloverde is all but leafless, but the tree will still look more or less naked even after it has acquired a fresh crop of leaves following the summer monsoon rains. A foot-hill paloverde's leaves are so small that they do little to disguise the tree's green-barked scaffolding.

I gave paloverde #17 its distinguishing number in 1980, during a study of male tarantula hawk wasps. For a couple of months in the spring, these big black wasps claim entire trees as their personal property, one male to a paloverde; they compete to be an owner of a tree in order to receive females when they come looking for a mate on a hilltop. Thus it is that sex motivates male tarantula hawks to take an interest in the beat-up little paloverdes growing in the Usery Mountains.

My motivation for an interest in the trees was more prosaic, in-volving the need to label the territories occupied by males, the better to keep records on the wasps. But because my study of taran-tula hawks kept going over many springs, I gradually formed an attachment to the trees for their own sake as I came to know them as individuals.

Year after year paloverde #17 seems not to have grown or changed at all but to retain the same size and shape that it had when I first hiked up the ridge to watch wasps fighting for its possession. But perhaps my impression of trees frozen in time is an illusion. After all, from one year to the next I have been barely aware of changes in the height and weight of my children. Now, however, both sons are adults, and they somehow have metamorphosed from small infants into individuals who are taller and heavier than I am, a forceful demonstration that with the passage of sufficient time, many small and barely noticeable changes can produce dra-matic transformations.

Just as I have a family photo album, so too I have built up a photo-graphic record of paloverde #17, with shots taken in different years and different seasons, different times of day, and under different weather conditions. As a result, I can check whether my impression of an invariant plant is an illusion or not. My photographs reveal that although paloverde #17 rarely looks exactly the same from one snapshot to the next, the most obvious differences are caused by seasonal changes in flower and leaf production or by changes in the sun's angle and cloud cover in the background.

Still I have only been following the fortunes of paloverde #17 for about one decade. A decade in the life of a mature paloverde may be a paltry 2 or 3 percent of its maximum lifespan, far too short a period to detect the full range of changes that it will be subject to. Just as one could hardly expect to detect much change in the appearance of a human over 2 or 3 percent of his adult life (less than 3 years), so too it might be unrealistic for an observer to expect change in a paloverde until say, forty or fifty years had passed. Few of us will have the desire or opportunity to observe the same plant or patch of desert for five decades. To say nothing of a century, or two. But that's what it might take to detect natural changes in paloverdes that grow with glacial slowness and live for centuries if they survive their childhood. As I monitor paloverde #17 in the years ahead, I suppose I will have to learn to appreciate the inevitability of change as well as the reassuring, if perhaps illusory, sameness of the little tree in a world that in most places seems to be coming apart right before my eyes.

Today in early June, paloverde #17 has a somewhat altered appearance thanks to this year's modest crop of beans. The "beans" are too short and lumpy to pass for string beans but there are dozens of them, despite a shortage of rain last winter. The rich green pods festoon some branches, warming up fast in the increasing heat of a genuine summer day.

The seeds within the pods have matured quickly. At first they were a paler green than the coat that contained them. But now that they have reached full size, they will begin to harden and turn brown and so will the drying seed pod. The fragile pod will then drop from the branch on which it grew, carrying one to four fully ripened seeds to the ground below. Even as the paloverde permits most of its fruits to fall to the ground, a few will retain their grip on the branch where they grew. These tenacious pods will soon take on a sun-bleached hue and their seeds will rattle in their dried husk containers when wind shakes the paloverde awake.

From past experience, I know that the fallen fruits will litter the ground beneath the tree that bore them for a short time only. In a matter of weeks, what were once unbroken pods will fracture, split and shatter. Soon the shell fragments will scatter to new resting places, crumbling beneath bursages and creosotes. In no time at all, the only sign of the paloverde's burst of production of fruits will be a handful of pale blotched pods still clinging to the tree in the searing heat of July.

In a good year, many hundreds or thousands of fallen pods practically seem to evaporate from underneath every paloverde in the desert and paloverdes are the commonest tree in the Sonoran Desert. In the sample studied by one of Forrest Shreve's academic descendants, the ecologist Joseph McAuliffe, 97 percent of the seeds were gone in just two weeks. Who is responsible for the rapid removal of so much paloverde produce?

McAuliffe discovered that harvest crews of desert rodents get to work each late spring night to take advantage of the temporary food bonanza provided by the fallen pods. Wood rats and pocket mice pop out of their burrows and scamper to the shelter of a paloverde. There they cut open the brown pods and remove the seeds. Because many more seeds are available than any one pocket mouse can consume in a night, the little rodents begin to set some aside for a later date. As their name suggests, pocket mice have pockets or cheek pouches designed expressly for the purpose of transporting seeds and other food items from one place to another. They and certain other rodents can stuff many food bits into their pouches, sometimes to the point of absurdity. But with faces swollen with future meals, they can efficiently cart quantities of food to their burrows or to food caches scattered about the areas in which they live.

Pocket mice are so diminutive, smaller than house mice, and paloverde seeds so hefty, that the mice usually accommodate just one seed per pouch. Nevertheless, the combined efforts of thousands of pocket mice and other seed-eaters serve to empty the fallen seed pods in short order.

Many of the uneaten seeds wind up buried in shallow caches a few inches deep beneath a bursage or other sheltering desert plant. Each cache contains one to more than a dozen seeds, depending on the species of rodent doing the food storing. Pocket mice typically bury one or two seeds per cache, thanks to their limited pouch space.

The goal of the foresighted rodent is to be able to come back and relocate the cache at a time when other food is in short supply. Some animals can remember the precise spot where they have stored food, returning to it in a moment of need. Certain forest birds like Clark's nutcracker and the pinyon jay, for example, have wonderful memories when it comes to food stores. They can keep in mind the exact spot where they buried a collection of pinyon nuts for weeks or months, as some elegant experiments have now shown. No one has conducted similar studies with pocket mice or kangaroo rats, but it would not surprise me if these food-caching rodents also have the ability to remember where it was that they placed a mouthful of food for safe-keeping.

Pocket mice do not necessarily have to keep track of the exact locations of their dozens of caches. McAuliffe and others have shown experimentally that various rodents can find buried seeds that they have not personally buried. The mice and rats perform this trick by detecting the faint odors emanating from seeds buried beneath several inches of soil. They use these cues to excavate the food, whether it was placed there by themselves or by a fellow rodent or by a human experimenter. Therefore, a pocket mouse need not possess an extraordinary map memory in order to find hidden food again but instead can sometimes locate old food caches by scent alone.

Whatever the basis for the discovery of food cached in June, a considerable number of seeds have not been found, excavated, and eaten by late July or August. When the rains come at this time of year, the undiscovered seeds have a chance to germinate and become baby paloverdes, rather than pocket mice chow. Thus, desert mice and other rodents unwittingly play an important role in the establishment of paloverde seedlings. First, by burying seeds, these animals play the role of Johnny Paloverde-seed, placing the seeds in position to germinate when soil moisture levels rise.

Second, in handling and nibbling on the seeds prior to burying them, they scarify the seeds in a way that eventually helps them germinate. Like many other exceptionally hard seeds, those of paloverdes are reluctant to absorb moisture unless the seed coat has been nicked or cut.

Third, by quickly removing fallen seeds, the desert mice and rats make it unprofitable for another seed consumer to bother with seeds after they have dropped from the trees. Primary among these other consumers are seed beetles, whose adult females lay eggs on the seed pods; the eggs hatch into larvae that cut their way through the pod and into a seed, which they devour. The female beetles will not lay their eggs on fallen pods, as McAuliffe found when he experimentally placed a set of perfectly viable, uneaten seed pods on the ground in cages with mesh fine enough to exclude rodents but coarse enough to allow the easy passage of seed beetles. The refusal of the beetles to exploit this resource probably stems from the extreme efficiency with which pocket mice remove, eat or bury seeds from fallen pods. Given the near certainty that a seed on the ground will be promptly eaten or promptly buried, the beetles may have almost nothing to gain by trying to lay their eggs on ground-level pods, and so they avoid them to concentrate on pods hanging from trees.

For their part, paloverdes encourage desert rodents to bury their seeds by producing them in overwhelming numbers in good years and by permitting the pods to drop to the ground as soon as the seeds are ripened. These attributes enable many seeds to get away from the lethal beetle seed predators and into the hands (or cheek pouches) of the merely semi-lethal rodent seed predators. At least with pocket mice and the like, a paloverde seed has a chance that it will wind up buried and forgotten.

If paloverdes had any choice in the matter, they would be especially partial to pocket mice because, as noted, these small rodents typically place just one or two seeds into a cache whereas the larger desert rats bury seeds in clusters. Clustered seeds may be more vulnerable to rediscovery for the following reason. When cached seeds germinate after the summer rains, they become especially easy to locate by cache-searching rodents, presumably because these seeds give off chemical by-products, thanks to a reactivated metabolism.

In one experiment, McAuliffe showed that searching rodents were more than twice as likely to discover germinating seeds than inert, nongerminating seeds. Seeds germinating in groups presumably produce more total odor, which should make them more vulnerable than single cached seeds to passing rodents. In digging down to reach the germinating seeds, the rodent may in passing excavate an already emerged seedling, killing it or at least reducing its chances for survival—even though pocket mice do not devour seedling paloverdes, only seeds.

The best of all possible worlds from the perspective of a paloverde would be to have its seeds distributed widely, one by one, in caches under desert shrubs. Then undiscovered seeds that germinated and became seedlings would not be inadvertently destroyed as a result of searches for other seeds in the cluster. Moreover, the seedling would be protected by the canopy of its "nurse" plant from the extremes of climate and from foraging rabbits, which do not turn up their noses at seedling paloverdes.

As it is, adult paloverdes do not have the best of all possible worlds to deal with when it comes to establishing their offspring. The rodents of the Sonoran Desert are in it for themselves, and any assistance they provide the tree is purely incidental to their own ends. Even so, one result of rodent maneuvers and paloverde seed production tactics is that a fair number of seeds germinate after a suitable rain or two and give rise to offspring that poke green cotyledons out of the ground. However, most infant paloverdes have short lives ahead of them. McAuliffe tagged several hundred seedlings that appeared after the summer rains in 1983. By the next summer, 92 percent of those growing in the open had been neatly snipped off at the base, probably by jackrabbits and cottontails. Thus, even after having successfully survived the obstacles to germination created by seed-sniffing rodents, conspicuous young paloverdes have almost no chance of living to their first birthday, let alone of reaching maturity.

But what about the saplings that had the good fortune to sprout beneath a bursage or other canopied plant? Largely hidden from view, these youngsters did better, as one might expect, although they were far from immune to foraging jackrabbits. Thirty-six percent of the concealed paloverdes survived into June, a rate four times higher than that experienced by those growing in the open. However, paloverdes sheltered by bursages are almost certainly competing for water with their larger companions. It would not be surprising if the struggle for water compromised the youngsters' growth and survival chances, although this point remains to be firmly established.

When I look at paloverde #17, a wonderfully mature oldster, I feel pleasure and even amazement, and rightly so. The seed that produced it ran a race with searching bruchid beetles and won. The fallen seed pod was probably harvested by a pocket mouse or other rodent, but the ancestor of paloverde #17 was not eaten outright. If and when it was buried in a cache, the seed was not relocated and consumed weeks or months after its burial. The seed managed to germinate and survive as a tender sapling, perhaps because it grew within the canopy of a bursage, now long deceased. For years it flirted with disaster but every time a jackrabbit wandered by, the rabbit overlooked the growing tree or at least did not destroy it entirely when it grazed upon the plant. Now the paloverde has become far too large to be nibbled to death by rabbits. The tree has conquered obstacle after obstacle, admittedly largely through good luck, but luck is worth cheering about and so is paloverde #17.

This year's seed crop is small, a fact that may enable the local rodents to consume every last seed. If true, this will not be the year that paloverde #17 generates a descendant to replace itself. But it has time, like the giant saguaro cacti nearby in the desert, and with more good luck, one of its seeds will give rise to its replacement some distance away before it dies. Perhaps the small paloverde growing on the ridgeline forty feet from paloverde #17 is its offspring, in which case it already has reproduced successfully, altering the Usery Mountain landscape for many years to come.

A family of black-tailed gnatcatchers flies from the smaller tree to paloverde #17 on a collective insect-collecting mission. The little birds slip from limb to limb in the now leafless tree, long black tails in constant motion, dark eyes searching for new victims, the birds' thin bodies full of life.

DESERT PEOPLE

But there have always been some . . . willful loners. And out alone
for a time yourself, you have some illusion of knowing why they
are as they are. You hear the big inhuman pulse they listen for, by
themselves, and you know their shy nausea around men and the
relief of escape. Or think you do.

JOHN GRAVES, *Goodbye to a River*

Schilling's best

From the top of Usery Peak, you can see the secret valley with its boulder-protected entrance lying far below between two ridges. The ridges break away from the main mountain and head south in parallel as they descend, before curling around to nearly meet far to the southwest. The approaching ridges fail to unite because the wash that drains the valley has cut a narrow rocky canyon between them before running out in the open where the wash offers itself up to off-roaders.

The two ridgelines both form and protect this pocket of desert, which is sheltered from more than just the vehicular manifestations of civilization, a civilization that has the capacity to impose such rapid and massive change on the desert that it threatens to overwhelm the natural evolutionary dynamic of the place. To the west behind one ridge, a mining operation has been at work in a half-hearted fashion for years. Gouged earth and a panoply of mining equipment pockmark the desert only a mile or so from the hidden valley. To the southeast of the other protective ridge, a spidery array of dirt tracks head toward the basin but are turned away at the last moment by the foothills barrier.

A huge golf course in the making reclines smugly on the desert still farther to the south. Bulldozers have smoothed out the would-be fairways, but for some reason, the brilliant green grass that someday will carpet the course has yet to be pampered into place. On winter days the gray-brown haze over Phoenix hides the skyscrapers and smudges a hundred miles of horizon. But in the encircled valley, the local saguaros and wolfberries, chollas and coyotes can pretend that the desert is as it was two hundred or five thousand years ago, a time when people were not the dominant feature of this world.

It is true that occasionally a thrown-away sheet from the *Arizona Republic*, wrinkled and yellowed from an outdoor life, blows into the concealed pocket and catches on a staghorn cactus, where it slowly fragments in the wind. Or a silvery balloon, constructed of paper-thin aluminum, floats in from an urban celebration elsewhere to spend the rest of its life deflated beneath an ironwood.

These latter-day artifacts make it more difficult to pretend that this isolated desert protectorate is as free from human affairs as it seems to be at first glance. But at least no human personally transported and deposited these items of debris directly into the Usery Mountain refuge. Instead, they relied on westerly winds to bring them from roads and towns that cannot be seen from within the valley.

Recently, however, I found two objects in this refuge whose discovery chipped away at the mental illusion of wilderness that I had created for the place. Far down the basin, just before the wash cuts deeper into the earth, carving out a staircase that descends among rock walls, a small piece of a block of salt lies half hidden among some tall *Ambrosia* stalks growing by the dry streambed. Someone, perhaps a rancher, perhaps a decade ago, had somehow carried in the once-hefty, although now much-diminished salt block, perhaps for his horses or cattle, which apparently occupied this isolated canyon then. The grazing permit for this part of the Tonto National Forest has long been retired, but the well-licked salt block will persist for a bit longer.

Some time after discovering the salt block I came across a partly buried metal cap about 4 inches in diameter, which poked out of the gravel on the side of the western ridge. The well-rusted but still intact cap had been engraved with a label—"Schilling's Best 2½ lbs." Oddly, the message appeared in reverse on the upper side of the cap to be read for what it was only when the cap was upside down, as it would be if a person inverted it after removing it from a jar or can.

The unusual method of labeling the cap and the weight of the material (2½ lbs.) that once filled the now absent container suggested to me that I had an antique, or at least a semi-antique, in hand. I intended to track down the date of the cap's manufacture but promptly misplaced my find and months elapsed without any detective work at all. Rediscovering the cap for the second time (in a closet at home), I quickly began to search out the history of the artifact before I lost it again.

The first thing I learned was that the Schilling's company no longer exists as an independent corporation because it was sold prior to 1947 to McCormick and Company, an East Coast firm dealing in spices. A call to their headquarters in Maryland resulted in a number of transfers, as one listener after another politely but quickly passed me on to another colleague. Finally, Mr. Jim Lyons of public relations came on the line to tell me that although there was a great deal of information available locally on antique McCormick food and spice containers, he could not help me identify old Schilling's containers. However, he gave me a California number for the Schilling's enterprise, which still does business but as part of the McCormick empire. The Schilling's public relations manager was baffled by my request, but she sent me on to a Mr. Jim Smith. He listened to my description and told me he'd be back in touch. To my surprise and pleasure, he did call back—within a matter of hours—to tell me that I had found a cap of a sort manufactured between 1941 and 1946. During World War II metal was in short supply and the cap was specially designed to seal glass jars. The jar that was once sealed by my find held 2½ pounds of coffee.

Just how or precisely when the Schilling's cap arrived in the hidden valley remains a mystery, although I like to think that the cap was stolen by a ringtail cat from a prospector's camp to be dropped in the desert by the presumably disappointed thief. Perhaps the actual history of the cap is less exciting, but its survival in the sheltered pocket behind Usery Peak testifies to the quality of Schilling's packaging and the astonishing abundance of people in the desert, who long ago occupied every secret valley of our world, dropping their trash behind them as they came and went.

The miner's cat

Pack it in, pack it out. There are no trash cans along the trails in the Superstition Mountains, and this is as it should be, particularly since much of the region is an officially designated Wilderness Area. Visible from the Userys, the Superstitions are much grander than my local walking ground. They offer a maze of broken mountains and canyons to hosts of hikers and backpackers, most of whom (but not all) obey the admonition to cart their garbage out with them when they leave.

My son Joe, his friend Paul Buseck, and I intend to be members of the responsible majority as we begin a one-night backpacking expedition on a gray January day. The introductory part of the hike consists of a four-wheel drive track, which we traverse by foot, plodding through long stretches of gluelike, yellow mud. A sullen sky spits a mixture of mist and snow flurries at us from time to time before we reach firmer footing at the start of a narrow canyon. We walk all day, along washes in canyons, hauling ourselves up hillsides and over ridges, in sunshine and under somber overcast, along obvious and not-so-obvious trails. A recent snow carpets much of the higher elevations but not so heavily that the trail disappears. In the early afternoon, air temperatures rise enough so that the snow begins to melt and rivulets flow down the trail, muddying our path again.

Wild clouds, a union of dark gray and snowy white, rush and swirl over the yellow cliffs on a mountainside to our right. A scrub jay squawks as it dives downhill on outstretched wings, the primaries curled up at the wingtips.

As the short winter's day concludes and thin clouds obscure the sun, we march on, stepping right over a little row of rocks that signal a doubling-back of the main trail. We will find the row tomorrow and interpret it properly. But for the moment we mistakenly follow a side trail that becomes faint, then fainter still. We persist, although increasingly convinced that we are off track, and succeed in following something like a trail to an overlook atop a cliff. From there we survey one of the main drainages of the Superstitions, a deep canyon a thousand feet or so below us. The wall on the distant western side of the drainage is the front range of the Superstitions. Snow covers the mountain ridge before us, which runs away to the south. We decide that we have reached our campsite.

Although we are not camped in the snow on our cliff top, the late afternoon is cold, and as soon as the sun slips beneath the mountains to our west, cold becomes colder. My son activates the little single-burner backpack "stove" and tosses noodles and canned chicken into a small pot on the heater. We eat in shifts, I with as much of my body as possible inserted in my sleeping bag where I experience an almost religious wave of gratitude for hot food, sleeping bags, and long johns.

My backpacker's sleeping pad, however, leaves something to be desired and never earns (nor deserves) my thanks. The night cold, which penetrates sleeping bag and multiple layers of clothes, and my inability to find a comfortable position on the pad combine to create a largely sleepless night. The clouds have moved off and a full panoply of wintry stars illuminates the sky above the Superstitions. I have ample opportunities to star gaze.

At some time in the middle of this endless silent night a sudden volley of loud rustles rouses all three of us from whatever dozing dreams we have managed to devise. With one or another flashlight in action we establish that the violent scrabbling is associated with our plastic garbage bag, which stores (for the return trip) the empty cans of chicken and other flotsam and jetsam from our evening meal. Although without my eyeglasses I am visually handicapped, I see that a squirrel-sized creature with a more than squirrel-sized tail has inserted its head inside the garbage bag. Frightened by our light beams and our comments, the thing bounds off—only to return on several other occasions to consume Paul's cinnamon rolls. Eventually, the animal decamps with the entire garbage bag as a final prize of the evening. On one of its visits, I see the creature well enough to establish that our visitor is a ringtail, a member of the raccoon family. Like the common raccoon, it sports a fluffy black and white ringed tail, but it is a far smaller and more delicate beast than its familiar relative.

After the ringtail departs, dragging the camp's garbage with it, silence reclaims the cliff top and I resume my wait for dawn, which is a long time coming. Finally, morning mercifully arrives and in the marginally warmer light of the new day I find and re-collect what remains of our garbage and our garbage bag. The various elements lie scattered about on a ledge just below our campsite, the cans licked clean, the scraps of food gone, the pieces of paper intact.

The behavior of our nocturnal guest was not at all unusual because the ringtail is a desert animal that has come to grips with people. At first glance, one would think that this species is an unlikely candidate to have made such an adjustment. A highly nocturnal creature, almost never seen during the day, ringtails occupy arid cliffs and rocky slopes in the western United States. They are solitary, each animal requiring many acres of rugged terrain as its private hunting preserve for lizards, rats, and even rabbits.

Despite their fondness for solitude, rugged habitat, and live prey (although they also consume some plant matter), ringtails speedily accommodate themselves and their diets to humans whenever possible, as the fate of Paul's cinnamon rolls vividly demonstrates. In fact, one other name for the ringtail is the miner's cat, so-called because of the vaguely catlike appearance of the animal and its willingness, even eagerness, to take up residence in prospector's camps. Like the common raccoon, which also thrives on the by-products of human activity, ringtails find human foodstuffs irresistible, whether in pristine form or as garbage. The first report on ringtail behavior, which appeared in the Journal of Zoology in 1842, focused on this aspect of the relationship between ringtail and man.

"The animal, Mr. Thomson has been informed, is found in most parts of the republic of Mexico, but is not known beyond the habitations of man. Besides fowls, butchers' meat, &c., it will eat bread, fruit and sugar; it breeds principally in outhouses, and particularly in neglected spots, producing three or four at birth. Sometimes it is tamed, and used like the domestic cat to destroy rats, mice, &c."

It was no doubt the fondness of ringtails for the rats and mice attracted to the living quarters of miners in the West that endeared them to these individuals. I wonder if "our" ringtail might not be descended from ancestors that profitably interacted with one or another of the many miners who have prospected in the Superstitions. These mountains, like most in the West, are riddled with prospectors' holes and abandoned mine shafts. Tailings from old mines stand out on the hillsides as nearly vegetation-free mounds, largely uncolonized after all these years.

The Superstitions have a special place in the mining history of Arizona because of their association with the Lost Dutchman Mine. According to local mythology, a miner named Jacob Waltz discovered gold in the Superstitions in the late nineteenth century but died without revealing the precise location of his shaft. It little matters that there is scant evidence that Mr. Waltz ever existed. Nor do many realize that the volcanic origins of the Superstitions all but preclude the existence of gold in these hills. All that counts is the rumor of gold, which has sufficed to propel a host of dreamers and greedheads into the Superstitions where they have rearranged portions of the landscape in an eager search for the lost gold. Sooner or later, most have left—perhaps to do meaningful work elsewhere.

As late as 1984, it was entirely legal to stake mining claims in the wildest and least man-altered parts of the Superstitions. Most of the considerable number of active claims at the time had been filed by people in the hopes of finding gold, either the Lost Dutchman or a new source. These persons operated under the Mining Law of 1872, a genuinely archaic law in every sense of the word but a law that remains the law of the land to this day.

The Mining Law of 1872 has as its simple guiding principle the notion that priority of access to public lands should be granted to those who can derive the greatest economic gain from these lands. To the extent that valuable minerals exist on public lands, they should be removed, according to the congressmen who devised the law in 1872. This position continues to find powerful advocates even now, which is why the Mining Law of 1872 has survived unchanged for well over a century. If you discover economically viable deposits of gold, lead, copper, zinc, cobalt, or other hard rock minerals on public lands, you can file a claim for them (assuming that someone else hasn't already beaten you to the punch). You will be charged as little as $2.50 per acre for the lands you wish to exploit. You will not be required to pay a royalty on the profits that you do secure (unlike those persons who extract coal, gas, or oil from public lands). While you are engaged in turning the land upside down in the search for mineral wealth, you may live on your claim holdings. You are unlikely to be required to restore the land to anything like its original configuration after completing your extractive enterprise. These arrangements appeal to hard rock miners and mining companies as a superb deal.

But what Congress giveth, Congress can take away. They have actually done so with respect to certain patches of public land over the years. One such area is the Superstition Wilderness Area, which was withdrawn from mineral entry by an act of Congress in 1984. At this time representatives of the government checked all the active mining claims in the Wilderness Area to determine whether there were any minerals of economic value in these areas. None of the roughly 20 claimants working in the Superstitions at this time actually had found an economically viable deposit. Nor was this situation at all unusual. Of 240 claims randomly selected from around the West, 239 were not being mined at the time of a recent inspection by the General Accounting Office. In fact, 237 had apparently never been mined. It is common practice for persons filing mining claims to do so speculatively, in the hope that a larger mining concern will eventually buy out their claims. Alternatively, many "miners" are individuals who wish to live or camp on public lands at public expense.

Those "prospectors" camped out more or less permanently in the Superstition Mountains had their claims declared invalid and were encouraged to leave for the sake of the environment. Apparently, none was forcibly evicted from the region. Nevertheless, thanks to congressional edict, no new mining claims have been accepted for the Superstition Wilderness Area since 1984, a small but measurable victory for common sense.

There is one exception to the prohibition on mining-related activity in the Superstitions. One may still apply for a treasure trove permit to hunt for the imaginary gold left in the Lost Dutchman Mine. Some groups have received said permit but have failed to find the treasure, to no one's special surprise. Yet perhaps, as the permit holders have searched through the Superstition Wilderness Area, they have provided a humble treasure or two for the resident ringtails, which are always ready to profit from their special symbiosis with the premier garbage-producing species of them all.

Where did all the glyptodonts go?

In the course of our walk through the Superstition backcountry, Paul, Joe and I came across very little modern trash, perhaps because of the blanket of snow that would have concealed any orange rinds and cellophane wrappers left behind by previous backpackers. In places where the snow had melted on the trail, however, we did see a considerable quantity of paleotrash, the stone chips and occasional potsherds left behind by much earlier inhabitants of the Superstitions, who were not recreational users of the landscape but permanent residents hundreds, even thousands, of years before we stepped onto the scene.

The human settlement of the Southwest is a story of repeated migrations of different peoples, each tiny group of successful pioneers amazingly fecund, resource-hungry, and trash-happy. Each wave of immigrants has altered the environment, inadvertently and advertently, a point that is obvious with respect to the current inhabitants of the Sonoran Desert. From my customary post on Usery Peak the signs of the most recent migrants, Anglo and Hispanic, are overwhelming: the sprawl of housing developments to the south and west; the brown photochemical haze hanging over downtown Phoenix thirty miles away; the white Central Arizona Project canal slinking through the desert for hundreds of miles, transporting water stolen from the Colorado River; the huge landfill on the Salt River Indian Reservation at the edge of the dry Salt River bed.

Closer to the Peak, the Salt River has water in it, courtesy of releases from Saguaro Lake authorized by the Salt River Project. The moving water is headed for the Granite Reef diversion dam that will intercept it and steer it into cement-lined canals for travel to treatment centers and then on to hundreds of thousands of households in greater Phoenix.

The Verde River joins the Salt right above the diversion dam. The water that ripples over the gravel and stone river bottom in the Verde passes the Fort McDowell Indian Reservation just a few miles upstream from Granite Reef. The reservation is home to a few hundred descendants of the Yavapai people who occupied central Arizona for hundreds of years prior to the arrival of the Spanish and then the Anglo migrants. The Yavapai's reservation was in turn once farmed by the Hohokam, a culture that evolved in the Sonoran Desert over a couple of thousand years before disappearing mysteriously in the 1400s. The durable trash of the Hohokam, their potsherds and the stone flakes discarded during tool-making, occupies thousands of sites among the Superstitions and along the borders of the Verde and the Salt rivers where smooth ocher-red fragments of broken pots, thrown away six hundred or a thousand years ago, lie half-exposed in mounds of dirt excavated by round-tailed ground squirrels.

The very earliest human occupants of the Southwest antedated even the Hohokam by several thousands of additional years. In fact, these migrants probably reached Arizona before the saguaros returned from their ice age refuge in Mexico ten or eleven thousand years before the present. They were far, far fewer than the current millions of desert dwellers, but even they doubtless produced and disposed of their fair share of garbage. However, the major environmental effects of the little bands of original pioneers may have been a good deal more dramatic than some modest littering. These first westerners have been accused of engineering a mass extinction of wildlife, erasing everything from glyptodonts to mastodons, an extinction unparalleled in human history until the present-day orgy of destruction taking place as tropical rain forests are converted into degraded pastures.

One thing is certain. There once were mammoths in North America and now there are none. As a result, mammoth dung is currently in short supply. But I have been lucky enough to have seen a mammoth dropping, thanks to Paul Martin, who has one squirreled away in a drawer in the Desert Laboratory where he works. What he showed me looked rather like a very large, very dry ball of horse dung deposited by an animal that had been on an extraordinarily high-fiber diet, the envy of modern breakfast cereal producers.

The mammoth deposit comes from Bechan Cave ("bechan" is a word derived from the Navajo discretely translated as "big feces") in southern Utah where it hit the cave floor about 12,000 years ago, according to radioactive dating techniques. That's a long time for dung of any sort to remain intact, but the specimen looks like it could be a couple of months old, instead of going on 12 millennia.

The salvation of Martin's souvenir of the late Pleistocene probably stemmed from the cool, dry conditions that prevailed in the cave that the mammoth used as shelter. The chemistry of the sand and rocks that covered the deposit may also have helped to preserve it. And in similar caves scattered throughout the West and elsewhere, other collections of "fossil" dung have persisted, left behind by animals now as extinct as mammoths and dead as dodos.

Among these cave-users was a menagerie of ground sloths, some of which were prolific in their contributions of dung that have survived to this day. Although not quite in the same ballpark as the elephantine mammoth, a five- to seven-ton animal, ground sloths were definitely not the ballerinas of the Old West. One species, the giant ground sloth, checked in at over three tons when mature.

Even the smaller sloths were capable of generating considerable waste. Two of the more famous sites that sheltered some of the lesser ground sloths are Rampart Cave in the Grand Canyon and Last Chance Cave in Tierra del Fuego. In both places, so many animals occupied the caves over so many years that the dung piled up six feet deep.

The deposits of these beasts have also been dated using the standard radioactive carbon technique. Interestingly, the most recent specimens from Arizona prove to be about 11,000 years old, not much younger than the mammoth dung in the Desert Laboratory collection and about the same age of the last mammoth remains to be found anywhere in the United States. Apparently, mammoths and ground sloths both bit the dust at right about the same time in North America, and so did toxodons, glyptodonts and gomphotheres, as well as a number of native horses, camels, and llamas, all of them big plant-eating mammals that are part of what is known collectively as the megafauna of the Americas.

Here we have a remarkable coincidence, the abrupt disappearance of a whole catalogue of hefty, often bizarre animals, a total of fifty-five extinct species of large mammals, enough variety to populate two continents with a diversity that matches the best Africa has to offer. What a thrill it would be to have a national park in the United States where you might see a glyptodont today, an overgrown, heavily armored armadillo weighing 1400 pounds, or where you could admire a ground sloth the size of a huge brown bear browsing calmly on creosote bushes. Instead we are left with just fifteen surviving big game species from the megafauna era.

We missed the chance to have Africa in America by a mere (in geologic terms) 10,000 years. Why did so many magnificent animals go extinct all at once?

Paul Martin has an answer. He blames human beings. He doesn't buy the idea that a climate change doomed the American megafauna, although it is true that a major shift in weather took place between 12,000 to 10,000 years ago. For some reason, temperatures climbed significantly during this period, causing the massive glaciers that dominated much of northern North America to retreat far back from the regions they had covered during their heyday. As the glaciers returned to the high arctic, southern Canada and the United States became warmer and drier, with woodlands and dry prairies replacing a habitat-type called the arctic steppe, a cold but rich grassland that once covered northern parts of the United States. In southern Arizona, the juniper woodland began its conversion to Sonoran desertscrub, and saguaros reentered the state around this time.

Advocates of the changing climate hypothesis argue that the environmental changes caused by the glaciers' retreat removed the habitats needed by the now extinct megafauna. Martin points out, however, that ice ages came and went several times in the Pleistocene, but massive extinctions without replacements by new ecologically equivalent species occurred only during the waning of the most recent period of glaciation. Furthermore, the extinctions that did take place at the end of the last glacial period fell disproportionately on the large plant-eating mammals of the period. Most of the small- and medium-sized herbivores (the rodents, the deer, the smaller peccaries) that roamed North America 12,000 years ago are still very much with us. The climate and habitats changed just as much for them as for the glyptodonts, but the glyptodonts are not here to entertain us whereas the smaller mule deer are.

Therefore, Martin asks (rhetorically) if we should put all our money on the climate hypothesis when there was another important change that occurred in North America about 12,000 years ago. The change that Martin has his eye on is the arrival of the first human immigrants to this continent. Although Africa, Europe and Asia have been home to people for tens of thousands of years and even Australia was colonized more than 30,000 years ago, the Americas did without us very nicely, thank you, until relatively recently.

Although there are fierce and continuing arguments on this matter, one widely accepted hypothesis on the colonization of America has the first people hiking over from Siberia to Alaska roughly

12,000 years ago. The descendants of these original North Americans then managed to move down through a glacier-free corridor from Alaska via British Columbia, making their way into what is now the northwestern United States in relatively few generations.

Even if some bands of humans preceded the invasion of 12,000 years ago, these forerunners appear to have been few and far between. Archaeological sites older than 12,000 years are extremely scarce in the Americas, and the dates associated with them are subject to ongoing dispute. In contrast, there is no argument that people were definitely here between 11,000 and 12,000 years before the present and that they crafted beautiful fluted spear points, some of which they placed forcefully between the ribs of mammoths, where they were discovered thousands of years later by modern archaeologists.

Paul Martin attributes extinction of the American megafauna to the makers of these spear points, a "tribe" now named the Clovis people. Martin argues that since most big-game species of the Americas evolved in continents free from people, they would not have been highly wary of humans. Similarly, on islands from which people have been absent until fairly recently, like the Galapagos, the locally evolved animals treat humans almost as if they do not exist. Such a blasé response almost invariably proves to be a mistake on their part, but when creatures live in environments without people, they have no opportunity to evolve the appropriate fearful response to the sight, sound and odors of humans. Unwary island fauna are sitting ducks for hunting people, when they arrive, although these animals react effectively to the appearance of one of their "natural" predators with which they have interacted over the millennia.

Since the glyptodonts and gomphotheres of the Americas evolved on island-continents free from humans, they, too, may have more or less ignored the Clovis hunters that they met for the first time eleven or twelve thousand years ago. If so, their failure to react would have been a mega-mistake, unavoidable given their history but devastating nonetheless. The giant ground sloths and their companions collided not with amateurs armed with primitive weapons but with Paleolithic pros, whose hunting skills and technology had been honed on the steppes of Siberia, where the game was undoubtedly more cautious, having co-evolved with humans for many thousands of years.

Thus, Paul Martin imagines the Clovis hunters arriving into what for them was a supermarket filled with meat waiting to be carted back to the campfire. With the living this easy, the human population would have exploded, and each new generation would have fanned out ever deeper into the Americas. Martin envisions a moving wave of humanity sweeping across first the North American continent and then down through South America, cleaning out the populations of vulnerable big game so quickly that the megafauna did not have time to adapt to the novel killers in their midst. Martin believes that a growth rate of about 3 percent per year (a rate well within the capacity of humans as modern Kenyans demonstrate today) would produce a population large enough to have extinguished most of the big game in North America in just 300 years.

Martin's hypothesis has many virtues, not the least of which is the dramatic story that it tells of humans descending like locusts on a new world and devastating what they found, a Paleolithic version of twentieth-century real estate developers. The melodrama of the hypothesis has drawn much attention to it, but the important question from a scientific perspective is, can we test Martin's idea rigorously?

Martin and others have indeed been able to suggest certain tests of the "overkill" hypothesis. First, the hypothesis would become as extinct as the ground sloths if someone could find deposits of mammoth or ground sloth bones in North America that reliably dated to less than 10,000 years before the present. So far no one has.

Second, Martin's hypothesis predicts that colonization of islands by human hunters should result in the extinction of big game, and the prediction is met in some cases. A notable example involves the Maoris and the moas of New Zealand. New Zealand was once people-free and was populated instead with a wacky zoo of creatures, including a gang of big flightless birds, the moas. Among the dozen or so species of moas was one genuine behemoth topping out at 10 feet high and weighing in at around 500 pounds. All of the moas were large birds; none weighed much less than 50 pounds.

The moas never met people until about 1,000 years ago, when a sea-faring tribe, the Maoris, colonized the place. In fairly short order, perhaps four or five centuries, the Maoris apparently killed every last one of the giant birds. Although the Maori role in the demise of the moas may have been partly indirect, with habitat alteration and the introduction of dogs and rats playing their roles, the Maoris clearly hunted moas with enthusiasm, judging from the thousands of moa bones that have now been recovered from Maori hearths. Jared Diamond and Paul Martin believe that the big birds were easy kills for Maori hunters, who probably were able to walk right up to them with snares, clubs and spears. As occupants in a land without people, the moas had never evolved an adaptive fear and loathing of humans.

Admittedly, there is a difference between the dimensions of New Zealand and North America. But even so, the selective nature of the extinctions in New Zealand is striking evidence in support of the overkill hypothesis as applied to the Americas. As mentioned already, the big edible species were the ones to go under, not the lesser animals, which would not have attracted the kind of enthusiastic attention of Paleolithic hunters out to get as much meat as they could in as short a time as possible.

The moa extinctions, however, took place in New Zealand, not the New World. Don Grayson has engaged in a more direct effort to test the overkill hypothesis by examining evidence from North America. He explored the bird extinctions that took place in the late Pleistocene of North America at about the time that the mammalian megafauna were disappearing from the face of the continent. Grayson recognized that if the overkill hypothesis applies to birds, we would expect unusually large, edible species to suffer disproportionate rates of extinction. If, however, the climatic change hypothesis were correct, a substantial number of smaller or inedible birds should have gone extinct by the end of the last glaciation, assuming as before that climate changes induce habitat changes that can make it impossible for certain species to persist.

In examining the fairly lengthy list of extinct birds from this period, Grayson found only one or two that might have been extinguished by hunters. Therefore, he concluded that the high extinction rate of Pleistocene birds suggests that something other than hunting pressure was responsible for their demise.

David Steadman and Paul Martin reviewed Grayson's evidence with a skeptical eye, noting first that although 60 percent of the megafaunal *genera* (groupings of related species) went extinct, only 10 to 20 percent of the avian genera disappeared at the same time. In other words, large mammals seemed to have suffered a greater probability of extinction than birds, as predicted by the overkill hypothesis. Moreover, the unhuntable birds that did go belly up included three large condorlike birds, a number of smaller vultures, a caracara, and a stork—all of which probably scavenged carcasses of dead mammals. These are precisely the kind of birds that would be vulnerable if the megamammals went extinct, taking their carcasses with them. Thus, it is entirely possible that overkill by Clovis hunters could have indirectly drawn a select host of avian scavengers into the pit of extinction without man ever having lifted a spear against these birds. Interestingly (and sadly), among the non-moas that disappeared from New Zealand shortly after humans arrived there was a truly gargantuan eagle, a thirty-pounder, far larger than any living bird of prey. This mega-eagle almost certainly preyed on moas and with their extinction, it, too, had a ticket to oblivion.

Although Grayson's challenge to Martin's hypothesis has perhaps been weakened by the rebuttal, the controversy around the overkill or blitzkrieg hypothesis still persists. Many paleontologists refuse to accept the notion that Paleoindians single-handedly demolished populations of big game that by Martin's own estimate totaled perhaps 100 million individuals at the time of the Clovis invasion. They point to the survival of the bison, which coped perfectly well, first with Paleoindians and more recently with bison-hunting cultures that replaced the Clovis people. As everyone knows, the American bison persisted in huge numbers right up until the end of the nineteenth century, at which time hunters armed with rifles nearly succeeded in sending the bison the way of the glyptodont. Martin notes, however, that the surviving bison was derived relatively recently from Asian stock, and so this species had evolved in an environment peopled with hunters. It had the opportunity to evolve wariness and fear of humans, abilities that many of the other big game species of North America may have lacked around 11,000 years ago.

Even so, the doubters generally prefer to discuss complex multifactorial hypotheses that attribute the disappearance of the megafauna to several interacting causes, with human predation just one possible factor. Martin comments that although these explanations may sound even-handed, they have a major disadvantage for practicing scientists, which is their resistance to a definitive test. In contrast, the overkill hypothesis lays it all on the line and can, as we have seen, be refuted cleanly when and if certain well-defined information comes to light. There is almost as much virtue in science in advancing an interesting idea that can be utterly dismissed, after testing, as in producing one that is supported by the evidence. The firm elimination of a wrong idea represents progress. However, the psychology of humans is such that we prefer to be right rather than wrong; therefore, I suspect that Martin hopes very much that news devastating to his hypothesis never does emerge from one or another study. Even if it does, he can take solace in the fact that there was nothing small about his explanation for the extinction of the giants that once roamed this land.

Thirty-eight Apaches

The gomphotheres are out of here and the Clovis Indians are no more, but there are still Indians in Arizona. The tribe that occupies the Fort McDowell Indian Reservation is sometimes called the Yavapai Apache. Their green farmlands are clearly visible from Usery Peak, but they look more symmetrical and more immaculate than they can possibly be, thanks to the great distance between the peaktop and the reservation. From here I cannot see the clutter of weeds growing along little irrigation canals that bring Verde River water to the fields, or the red steers that wander into the neat green rectangles from the untidy and irregular mesquite groves that border the river.

Actually, to add the name "Apache" to the Yavapai is not historically sound. These people have a language that is totally different from that of the Apaches and an ancestry that links them with other non-Apachean tribes that occupy western Arizona. However, those Yavapai living in the nineteenth century on the eastern edge of their range did come in contact with genuine Apaches. They were friendly with them and intermarried to some extent, providing the beginnings of a connection. During the wars with the Anglos that broke out in the latter part of the nineteenth century, all the Yavapai and Apache were lumped together indiscriminately by the newly arrived settlers. The white pioneers not only called both groups Apaches, they treated them with equal hostility. Upon conclusion of the several years of warfare, the surviving Yavapai were rounded up and shipped off to the San Carlos Apache Indian Reservation, where they lived together with Apaches and intermarried freely with them. By the time the Yavapai were permitted to return to a reservation on the Verde River, they brought with them any number of Apache spouses. By this time, the Yavapai were more or less integrated into Apache lineages, making it possible in the end to justify the misnomer applied to them by the new immigrants to Arizona who so thoroughly displaced both Yavapai and Apache from their ancestral homes.

The Anglo settlers of Arizona attached the name Apache to many things besides the Yavapai Indians. Apache Wells. Apache Pass. Fort Apache. Apache Butte. Apache Lake. Apache Junction. The Apaches are remembered here. There is even a town called Geronimo in the center of the San Carlos Apache Indian Reservation (and a Geronimo Smokehouse on the western edge of the same reservation where you can buy cigarettes without the aggravation of a federally imposed excise tax). The town of Geronimo now consists of two or three buildings, all seemingly attached to an abandoned garage and store that once offered passersby "Cold Beer." The moribund establishment with its boarded windows now turns a cold shoulder to travelers, but a dusty neon "Budweiser" sign still looks out wistfully on the highway. On the wall of the ghost store a campaign poster urges us to vote for Bill J. Hawkins, County Supervisor, Democrat.

The town sign that commemorates Geronimo is a simple green rectangle with white lettering set on the verge of the road. In the summer, trailer trucks send the hot wind swirling among the lower branches of the big cottonwood and salt cedars that stand near the sign. A half mile down the road a historical marker announces that the original Fort Thomas stood nearby in 1876 when it was built "to keep Geronimo's tribesmen on their farmlands along the Gila River."

The Apaches of Geronimo's day did not think of themselves as farmers or even as Apaches. The name they are widely known by is apparently derived from a Zuni word ("apachu") for "enemy." The enemies of the Zuni called themselves "Dine" or "N'de" or "Na-dene," words that mean "people" to the Apaches, who were warriors and raiders, not dusty cultivators.

One of the quintessential tribes of American folklore along with the Iroquois, the Sioux and the Navajo, it is hard to imagine the Southwest without them. But they, like so many other peoples, immigrated to Arizona in the not so distant past. Most archaeologists place the arrival of the first bands here sometime in the 1500s, thousands of years after the Clovis people came, slaughtered mammoths and (maybe) polished off the ground sloths before they themselves disappeared, probably through cultural change.

Linguists put the Apache and Navajo languages close together in a family of languages called the Athapaskan or Na-dene. The original Na-dene speakers occupied Alaska and the Yukon where some Indian groups retain the unique features of this language group. Linguistic theory suggests that the pioneer Na-dene arrived in Alaska about 2,000 years after the first Paleoindians migrated from Siberia to North America, roughly 12,000 years ago.

From the descendants of the original Na-dene came adventurous, tough or desperate bands that over the centuries traveled thousands of miles from their ancestors' new homelands, carrying their linguistic heritage with them. One set of immigrants settled on the coast of the Pacific Northwest; others went much farther, eventually reaching the Southwest where they multiplied and fractionated into the Navajo and Apache tribes, which in turn formed bands and semi-bands that fanned out over the land. In time the Apache came to occupy much of western Texas, New Mexico, and eastern Arizona. Among these Na-dene speakers were the Mescaleros, Jicarillas, Chiricahuas, Mimbres, Coyoteros, and White Mountain Apaches.

When the Apaches and Navajos showed up in the Southwest, the Zunis and Hopis and other Puebloan cultures had been here for centuries already, building their elaborate apartment dwellings, cultivating their fields of corn, squash and beans. These people have been placed in a different linguistic group altogether from the Navajo and Apache. Furthermore, key genetic similarities link them with most other Indian tribes of the Americas, not the Na-dene. The congruence between linguistic and genetic evidence supports the hypothesis that most American Indians descended from that one small group of closely related Paleoindians that marched into the New World 12,000 years ago (although there is debate about the date). This pioneering band gave rise to the Clovis culture and then hundreds more, all of which shared their ancestors' special genes and elements of their ancestors' language as well.

The Zunis and Hopis did not welcome the new nomadic immigrants, who had come a long way to find a new home and new peoples to fight. The Apaches were hunters and gatherers, many of whom engaged in rustling and raiding, although the various subgroups operated almost completely independently of each other and had different personalities and economic strategies. But stealing from others was part of making a living for many Apache groups, and they were good at it (although not as proficient in this regard as the many Anglo settlers who eventually deprived the Apaches of almost all their traditional homelands).

By the time Geronimo, an Apache associated with the Chiricahua Mountains of southeastern Arizona, was making a legend of himself, the Apaches had been through a great many changes, with more to come. The horses that the Apaches rode on their raids into Sonora, Mexico, were the distant descendants of horses introduced by the Spaniards sometime in the sixteenth century. The Apaches appear to have acquired their horses in the mid to early 1600s, probably by trading or by thievery from their neighbors.

The first Apaches into Arizona had no idea what a horse was all about, the native North American horses having gone the way of the glyptodonts around 10,000 years previously. But the Apaches quickly developed a full appreciation for the utility of a horse. The first certain report of a mounted Apache raid dates to 1659. On their long-distance travels, they fashioned "horseshoes" out of pads of leather, which they tied to their animals' feet to protect them against the battering of the journey.

When the Apaches met the first immigrants from the eastern United States, they had already had a substantial history of interactions with intruders into what they considered their territory. By the early 1800s Mexican miners and settlers had pushed far enough north into Sonora and Chihuahua to offer new opportunities for trading and raiding for the Apache bands that roamed through these areas. Not all their encounters degenerated into hostilities, but conflicts were common enough and major raids so frequent and effective that Mexican government officials in Sonora and Chihuahua embarked on an openly genocidal program designed to eliminate the Apache threat, or so they hoped. These officials offered bounties for the scalps of Indians, payments that were eventually set at 100 pesos for an adult male, 50 pesos for an adult female, and 25 pesos for a child. This offer and others like it attracted a certain kind of person, one of whom was an enterprising American named John Johnson who was in northern Mexico in 1837.

Johnson made his way north to the Animas Mountains in what is now southwestern New Mexico, but was then part of Mexico. He invited an Apache band led by Juan José Compa to come to trade with him at Agua Fria. On the third day of trading, Johnson and his confederates opened fire on his now unsuspecting clients with what has variously been described as a cannon or howitzer. Many Apaches died and many others were wounded.

Johnson collected his scalps and quickly made his exit from town. News of the atrocity quickly reached many Apaches, including a formidable fighter whom the Mexicans called Mangas Coloradas. This man understandably took offense when he learned what had happened in Agua Fria. Neither Mangas Coloradas nor any other Apache succeeded in capturing and killing the American organizer of the outrage; he went on to live an undeservedly long life in Mexico. Instead, the Apaches revenged themselves on Mexicans living in various parts of the Southwest. Their attacks eventually caused Mexican miners to abandon the town of Santa Rita well to the north of the Animas Mountains and to evacuate a large area in adjacent Sonora as well. During the campaign, Mangas Coloradas enlisted the assistance of other Apache bands. He and his confederates learned that they could preserve their lands from outside threats. Or so they thought.

As more Americans began to filter into the Southwest on trapping and mining expeditions, they ran into Apaches who, surprisingly enough, were inclined to treat them well, despite their knowledge of the massacre at Agua Fria. The Apaches apparently made a clear distinction between Mexicans and Americans, and initially there was a certain amount of cooperation between the two cultures. But history then repeated itself—up to a point. At first, the Apaches and the new immigrants traded warily with each other, but soon assorted killings took place, and then the two sides went at each other's throats.

The Civil War era was the major turning point in Apache-American relations. The United States had secured jurisdiction over much of Apache country from Mexico following the treaty that ended the Mexican War of 1846–1848 and also as a result of the Gadsden Purchase in 1854. Subsequently, American settlers poured into the Southwest, especially when gold was discovered in California and western New Mexico. These newcomers severely depleted the wild game in Apache lands and made existence difficult for the Indians, who numbered only a few thousand despite occupying a huge chunk of real estate. In late 1860, Mangas Coloradas, who by then was in his sixties and an Apache greatly respected by his tribesmen, could see the handwriting on the wall, which was that the Apaches were about to drown in a sea of white settlers. Recognizing the fanatic passion that Americans had for gold, he attempted to convince a group of miners that there was gold in Sonora, Mexico (which there was). His goal may have been to start a new gold rush that would induce the hordes of miners on Apache land to leave the area and head to Mexico. The miners were suspicious of Mangas Coloradas, and they decided to teach him a lesson. They did so by tying him up and then whipping him severely, apparently primarily for the pleasure of humiliating a savage. His mind clarified about the nature of these new invaders, Mangas Coloradas embarked on yet another campaign to clear the entire lot from his traditional homeland. He did not succeed as he had in his earlier battles against the Mexicans, but a great many people died on both sides.

Around this time, another major figure among another Apache group, Cochise, who now has a county in Arizona named after him, also learned the hard way that the new Anglos were his enemies. He did so during an encounter with an inexperienced Army officer, a Lt. George N. Bascom, who was part of the force sent to Apache territory to protect American settlers there. Bascom attempted to imprison Cochise and some of his relatives following a supposedly peaceful meeting to determine who had kidnapped the Hispanic stepson of a local rancher named John Ward. As it turned out, Cochise and his allies had nothing to do with the kidnapping, but Bascom was convinced that all Apaches were in league with each other and so decided to treat Cochise as the responsible party. Cochise dashed away from the tent in which the meeting took place, but a number of his fellow Apaches failed to escape with him. Wounded in his escape, Cochise then took a number of Americans hostage to exchange for the Apaches held prisoner by Bascom. Bascom refused the deal. Cochise had his captives executed. Bascom did the same for the Apaches in his hands. It was an unhappy time all round.

The kidnapped youngster survived and was reared by his Apache captors under the odd name Mickey Free. When his relatives came years later to retrieve him on the reservation to which he and his Apache foster parents had been sent, Mickey refused to leave. His relatives were baffled.

From 1861 on it was more or less continuous guerrilla war with the greatly outnumbered Apaches up against the U.S. Army and assorted civilian militias and gangs. By 1863 Mangas Coloradas probably could see who was going to win and may have realized that there was nothing he could do about it. In this year a group of miners approached his camp under the guise of the white flag of truce, only to capture the Apache leader and transport him to a military post, where that same night he was shot as he was "trying to escape." One account of his death is that his guards built a large fire near the elderly Apache and heated their bayonets before placing them on the feet and legs of the captive. Mangas Coloradas protested and moved to rise. In response, his captors riddled him with bullets from their muskets and six-shooters.

Later the Army doctor at Fort McLane, a Capt. D.B. Sturgeon, exhumed the body of the slain and scalped Apache from its shallow grave and removed the head of the corpse. Mangas Coloradas was a big man, 6' 4" tall, and the phrenologist, Professor O. S. Fowler, who received his skull discovered it to be larger than the skull of Daniel Webster. An official Army inquiry into the conditions surrounding the death of Mangas Coloradas absolved General Joseph R. West of any wrongdoing in the case.

The combination of continuous pressure from the Army and the influx of new settlers soon forced the various Apache bands to surrender and to accept reservation life. Cochise and the Chiricahua Apaches knuckled under in 1872, after being guaranteed freedom to roam within a reservation established for them in their traditional territory of the Chiricahua and Dragoon Mountains.

The reservation endured for four years, during which time the Apaches had an unusually sympathetic federal agent in charge, Tom Jeffords. In the 1870s, typical sentiment about the Apaches was encapsulated in the advice given by Samuel W. Cozzens, an Easterner who visited Arizona for an extended tour in 1858–1860 and who wrote a lengthy travelogue entitled The Marvellous Country; or, Three Years in Arizona and New Mexico, the Apaches' Home. Cozzens claimed that the Apaches "can never be subdued; they must be exterminated; and the sooner the American people realize this fact and act accordingly, the sooner will the fertile valleys of Arizona again wave with golden grain, her grazing lands be covered with ten thousand herds of cattle, . . . and the smoke ascend from the settler's happy home." Cozzens himself contributed to his suggested goal by killing several Apaches during his journeys in Apacheria, if his stories can be believed. By his own account, one of his victims posed no threat whatsoever to him or his party.

Jeffords did not share Cozzens's view of the matter, but he had to deal with the penny-pinching attitude of his superiors in the Indian Bureau and the restlessness of many Apaches on the reservation. Various bands regularly bolted for raids into Mexico, a strictly verboten activity from the American perspective, although a highly traditional pastime of the Chiricahua Apaches.

Eventually, the various pressures caused Jeffords to lose control of his charges on the reservation. With Arizonans screaming that something had to be done about the Indians, Jeffords was relieved of his post, the reservation was disbanded and the Army sent to round up the Apaches for shipment elsewhere. Many Chiricahua Apaches departed for Mexico, where some managed to avoid consignment to a reservation well into the 1930s.

After 1876, Geronimo bounced around reservations in New Mexico and Arizona with repeated "escapes" from federal control and unauthorized forays into Old Mexico. Although Geronimo was born in what is now southeastern Arizona, he knew well the Sierra Madre mountains to the south. His famous name is the one the Mexicans gave him; the English version of Geronimo is Jerome, hardly the kind of name that could inspire a myth-maker. His Apache name was Goyahkla (one who yawns). His activities as a free man raised alarms throughout southern Arizona and northern Mexico. The efforts to track him down and subdue him galvanized newspaper readers throughout the United States of America.

Finally, in 1886 Geronimo surrendered for the last time in Skeleton Canyon, just across the New Mexico–Arizona border in the Peloncillo Mountains, where he had been tracked down by troops under General Nelson Miles. The troops were the Apache Scouts, most of them Chiricahuas. Mickey Free, the kidnapped kid, served as an interpreter with these Army-employed Indian warriors. General Miles's predecessor, General George Crook, strongly advocated using armed Apaches to go after the renegade bands, advice that he put into practice himself. There was little or no solidarity among the various tribes and sub-tribes, and plenty of Apache men were glad to go hunting other Apaches, a task they performed with vastly greater skill and endurance than regular army forces.

When Geronimo agreed to give it up for good, he was accompanied by just thirty-seven people, according to one count. (The Anglo population of Arizona in the 1890 census, not long after the removal of Geronimo and his thirty-seven compatriots, was a little less than 60,000.) The captured Apaches were hurried over to the closest army post, Fort Bowie, and immediately shipped out to the nearest train station. The Fort Bowie Army Band sarcastically played "Auld Lang Syne" as the Indians were sent off. Barely avoiding a lynching party at a train stop early in their journey, the prisoners-of-war eventually traveled all the way to Fort Marion, Florida. Later they moved on to Mt. Vernon Barracks near Mobile, Alabama, and ultimately to Fort Sill, Oklahoma. Geronimo died at Fort Sill in 1909, still a prisoner-of-war twenty three years after his final surrender. Three years later, the remaining Chiricahuas were declared nonprisoners, and those who wished to do so were permitted to resettle on the Mescalero Reservation in New Mexico.

For the captured Apaches locked in their reservations, often in environments totally foreign to them, there was no realistic hope of turning back the clock, of erasing all the changes they had been through. But the thought crossed their minds. Like members of many other captive Indian populations elsewhere in the United States, they desperately wanted to believe that what had happened to them was just a bad dream. And in the period from 1880 to 1917, three men on Apache reservations developed sizable followings after prophesying a return to the old ways, a time when the white men would disappear and the Na-Dene would return to the lands that were rightfully theirs.

One after another the cults collapsed as their leaders were killed or key prophecies failed to materialize and the faithful dispersed, taking their disappointments with them. One of the cult organizers, Daslahdn, aware of the flagging enthusiasm of his converts, who had been participating in elaborate ceremonies for several years without the desired effect, insisted that his followers cut off his head—for he would return from the dead in three days as proof positive of the legitimacy of his prophecies. The cult obliged him in his macabre request and waited grimly for three days, after which time Daslahdn failed to revive. His disenchanted adherents probably realized then what was in store for them for the rest of their lives.

I drove out one August day to Skeleton Canyon on a dirt road that runs off the highway between Rodeo and Douglas and winds through a series of ranches before ending at a ranch house at the mouth of the canyon. A monsoon storm came sweeping up the valley from Mexico as I approached the end of the road. Black streamers of rain and bolts of lightning and I came together at the ranch house. Two large wet dogs bounded through the downpour toward the gate across the road. Cows stood hunched under some scabby mesquites in an overgrazed pasture. That was as close as I was to come to the place where Geronimo called it quits. The dogs continued to bark as I drove away. The washes flowed across the road as I retraced my journey through the land that the Chiricahua Apaches must have dreamed about during their long years of exile in places without canyons or mountains.

The last Indian war?

Although the Chiricahuas were home to Apaches for many years, these hunter-gatherers and raiders did not build elaborate stone structures, and so they left only subtle marks on the mountains and canyons that they occupied in southeastern Arizona and northern Mexico. But there are many places in the American West where Indian cultures altered their environment in ways that are apparent today, even to a casual observer. One of the best places to appreciate the mark of prehistoric peoples on the land is southeastern Utah, the domain of the Anasazi cliff-dwellers during the same period when the Hohokam irrigated squash, beans and corn on the floodplain of the Salt River near Usery Peak.

Much of the Anasazi's ancient homeland in southeastern Utah is now regulated by the Bureau of Land Management, which properly insists that hikers pick up free permits before heading off into Anasazi-land. For example, to hike into the Grand Gulch Primitive Area, one must first stop at the BLM's Ranger Station at Kane Gulch. I arrived there after closing time (4:30 P.M.), but finding the door still open (it was not yet 5:00), I entered and imposed on the bearded BLM ranger. He presented me with the necessary form and a sheet detailing how to behave around Anasazi ruins. He told me in particular to avoid walking on the middens by the cliff dwellings. I said I would. He queried, "Do you know what a midden is?" Instantly, I was back in the fourth grade faced with a severe Mrs. Walker, her wispy gray hair tied back in a bun, her paddle (the size of a cricket bat, if my memory is to be trusted) hanging in the cloakroom. I said, "I do." He persisted, "What is it?" I told him that a midden contained the remains of the garbage tossed out of the dwellings by their inhabitants; I assured him that I was in complete agreement with the BLM's policy of protecting the sites. He looked unconvinced but gave me my permit and I escaped without further interrogation.

Employees of the BLM have their hands full. There is one paid ranger to cover about 300,000 acres of Grand Gulch for six months of the year plus a handful of volunteers in the summer. There are hundreds of archaeological sites in the very rugged canyon and no way that they can be adequately policed. On one day in a recent spring 140 people registered to enter Grand Gulch via the Kane Gulch trailhead.

The next morning I made an early start, the only person that day to tramp off through the sagebrush flats to the canyon, following a trail that began gently enough, paralleling a little dry streambed cutting through low rock walls, walls that grew and grew as the watercourse descended. And descend it did, with vast sandstone walls soaring overhead, more and more monumental at every turn in the gulch.

A prairie falcon dashed across the ribbon of blue sky visible from the bottom of the canyon. Then, about four miles downhill, the ribbon expanded as a side canyon joined the main one, now called Grand Gulch. On the red-brown wall in front of me on an inaccessible ledge, a line of largely intact Anasazi dwellings looked out over the cottonwoods at the canyon junction. Their black doorways stared back at me.

Below the upper row of apartments, a matching set of small rooms rested on the ground of the alcove, the buildings largely destroyed by past pot hunters. A huge midden of sand, potsherds, stone flakes, tiny corncobs, bits of wood and other debris formed an apron below the old houses. Other visitors in the past had collected bits and pieces that appealed to them and placed the fragments on large rocks in and around the midden. I obeyed the BLM ranger's instructions and kept my feet off the prehistoric landfill and my hands to myself.

The Anasazis of the 1200s knew how to pick stunning places in which to live and dispose of their garbage. The canyon junction is now a favorite with backpackers, who during peak periods in the spring form a little village here containing as many or more people than lived at the Anasazi dwellings during their prime. But on the day of my visit, no one was camping beneath the giant cottonwoods at the confluence of the canyons. Only rufous hummingbirds traveled among the patches of Indian paintbrush along the streambed.

Over the next two miles, I found one ruin after another in alcoves on southfacing canyon walls. One major site in a huge pocket in Navajo sandstone contained a little "corral" of stick and wattle whose function puzzled me until I later read the name "Turkey Pen Ruin" on a BLM brochure about the place. The Anasazi domesticated turkeys and made elaborate cloaks from the plumage of these birds.

Most of the Grand Gulch ruins have no name, but each has its own personality. One small site with just a single intact building stood on a ledge far above the meandering stream. The stones used to construct the building still fit together to perfection. From the angled doorway, the Anasazi occupants of the house had a magnificent view down the twisting canyon, filled with silver and green-leaved cottonwoods.

About noon I heard the first thunder in the upper canyon and saw big cumulus clouds building up. Having walked as far as was comfortable, I began to retrace my steps, keeping an eye on the developing clouds. Thunder became more frequent, and my imagination filled the canyon with a wall of red water that would either sweep me downstream to my death or chase me from the streambed up onto the canyonside where I would be exposed to rain and lightning. My pace quickened accordingly, but it still required a long time to climb out of Kane Gulch, during which time I did not encounter a flash flood or even a shower. The storms I heard around noon had drifted off to the north, producing thin streamers of rain that reached down to barely touch the sweet-smelling sage flats.

The Anasazi, like the Hohokam, abandoned their homes in the late 1400s, leaving the land unoccupied for a time. No one knows precisely why they left or where they went. But in time new colonists from other cultures came to settle among the canyons and mesas of this desert land. They saw the cliff-dwellings preserved in the austere climate of the place and wondered about their absent occupants. None of the more recently arrived Indians became as famous as the Chiricahua Apache, Geronimo, although there were a good many Utes and Paiutes present during the Anglo invasion of the region. Southeastern Utah does claim, however, to have been the scene of the last Indian war, a claim that appears on a roadside plaque near Blanding, Utah.

I discover the plaque after my hike into the Grand Gulch Primitive Area, while on my way back to Arizona and a far more sedentary life. My itinerary takes me through Blanding, a small town founded by Mormon settlers who began their conquest of southeastern Utah in the late nineteenth century. In preparation for the drive, I checked my map and noticed a conspicuous symbol labeled "Chief Posey's War." The symbol represents a historical marker, which is located a short distance south of Blanding. With luck, I manage to find it on a sandy mound several hundred yards in from the main highway. Beside the dirt road that leads to it, the marker appears far less conspicuous than its representation on my map.

A smattering of trash, including one car tire, litters the roadside around the site. One stunted sunflower blooms near an empty paper cup. A complex blend of aromas from juniper, pinyon and sage issues from a shallow canyon that slopes off to the right.

Although I had thought that the last Indian war in the United States might be the conflict that ended with Geronimo's surrender in Skeleton Canyon, or perhaps the massacre of the Lakota Sioux at Wounded Knee that took place in 1890, the marker tells me that

Chief Posey's War occurred from March 20 to March 23, 1923. On the twentieth, according to the metal text, Chief Posey and some Paiutes loyal to him descended on Blanding to spring two tribesmen from the Blanding Jail. At a spot about ten miles south of town, site of the current marker, Posey shot the horse out from under one of his pursuers, John Rogers. This slowed pursuit for a time, and the Paiutes slipped off to Comb Ridge another fifteen miles or so to the west. But the next day Posey was tracked down and killed, as was one other Paiute man, and by the twenty-third all the participants in the raid had been rounded up and brought to trial. Their sentences were not given on the marker.

About ten miles farther down the road I pass through the White Mesa Ute Reservation, which has some government issue housing and some of local design perched along the highway. The occupants of these houses probably hold to a somewhat different version of Chief Posey's war than the one inscribed on the metal marker up the road, judging from an account assembled by the Utah historian Robert McPherson.

His reconstruction of events begins with the arrest of two Indian men, Joe Bishop's Little Boy and Sanup's Boy. The Paiutes had been apprehended by Sheriff William Oliver after they had robbed a sheep camp, killed a calf and burnt a bridge. After several days in custody, their trial took place on the twentieth. At noon, the trial went into recess for lunch, the two Paiutes having been already convicted but not yet sentenced, an event scheduled for the afternoon. The Sheriff was in the process of getting the convicted men out of the schoolhouse where the trial took place and back to jail when a scuffle erupted, apparently initiated by Joe Bishop's Little Boy. In the fight that followed, the Indian grabbed the sheriff's pistol, wounded the sheriff's horse and left town promptly on his own horse. Sanup's Boy and Chief Posey also hightailed it out of Blanding back to the Indian camp.

Many Paiutes were in town for the trial, and Sheriff Oliver and his fellow citizens rounded them all up. First they used the schoolhouse as a detention center for the Paiutes, but they later constructed a 100 foot × 100 foot stockade enclosed with barbed wire to hold the many detainees. Local Navajos were commissioned to build two hogans for shelter within the stockade.

After securing the bewildered Paiutes in the schoolhouse, a posse assembled and pursued those Indians that had joined Chief Posey and the two escapees. The two sides met and exchanged gunfire in the afternoon of March 20. Posey possessed a 30.06 rifle, which he used with moderately good effect, mortally wounding the horse of Deputy Sheriff John Rogers. But during the running gun battle, Posey himself was hit by Dave Black, although this fact was not known to the posse.

The Blanding contingent retired from the field and the entire town gathered that evening to discuss the day's adventures and to decide on a course of action. According to John Rogers, who considered himself lucky to be present at the meeting, "It was unanimously decided that this was going to be a fight to the finish. We all knew that Old Posey was not going to be taken alive. . . ."

The next day (March 21) the posse returned to the chase, locating some Indians near Comb Ridge and killing Joe Bishop's Little Boy. Afterwards, there was no more fight left in the Indians. The surviving Paiutes kept on the run for two more days before surrendering fearfully on March 23. Chief Posey was, however, not among the captives.

The Sheriff and his posse continued to search for Posey for almost a month. Finally, the Indians contacted a federal marshal, J. Ray Ward, and agreed to show him where Posey was if he promised not to tell the local Mormons of the burial site. He assented and the Indians took him to the grave of Chief Posey, who had slowly died of blood poisoning as a result of the gunshot wound he had received at the start of the "war."

Mr. Ward kept his word and urged that the people of Blanding be satisfied with the knowledge that Posey was definitely dead and buried. But the next day an expedition from Blanding tracked the marshal's path to the grave of Chief Posey. The Blanding men exhumed the body and photographed themselves in the company of the corpse.

Back in town, the young children of the Paiutes were forcibly separated from their families for a haircut, sponge bath, and a new set of clothes, after which they were shipped out to the Indian school at Towaoc, a native word that means "all right" or "just fine." The federal authorities finally settled on a little reservation of about 8500 acres for the Paiutes who had been corralled in Blanding. They permitted the adults to leave their miniature concentration camp for these lands, although the Indians made other moves in the 1940s and 1950s to the current White Mesa Ute Reservation.

As Robert McPherson points out, the "war" was really more a white affair than an Indian one. It was the Mormon community who decided to use the flight of the two Paiute prisoners as the excuse to resolve, once and for all, the problems caused by a handful of Indian neighbors. The events of March 1923 had been preceded by years of conflict between the Mormon settlers and the Paiutes onto whose lands they had brought their cattle and created their farms. Chief Posey had long been an especial irritant to the Mormons, with his demands for food and aggressive thievery from the settlers. Posey made it clear that he resented the presence of whites in his corner of Utah.

From the Indian perspective, Posey and his Paiute followers had every reason to be resentful. The federal government had, without consultation with the Paiutes and without making sufficient provision for their welfare, opened their traditional lands to Mormon settlement. These newcomers occupied the prime camping areas that had long been occupied by Paiutes. The town of Blanding usurped one of these areas. The Mormons brought cows, and plenty of them, to the region, and these animals changed the environment in ways harmful to the native game that the Paiutes depended upon.

Some of the Mormons knew full well that the Indians had suffered and were continuing to suffer, even starve, as a result of the deterioration of their ancestral hunting and gathering grounds. But for the majority, themselves barely scratching out a livelihood in difficult terrain, the loss of livestock to Indian rustlers and the demands for flour from these unwelcome neighbors created hatred and the desire to be rid of the Paiute altogether. Albert R. Lyman's book *Outlaws and Indians* presents the Mormon perspective on the Indian problem. Lyman was one of the original band of Mormon pioneers who entered what is now San Juan County in 1880, and he was the first to settle in Blanding (in 1905 with his wife Mary Ellen). Lyman describes Chief Posey on the one hand as not inherently an evil person ("His is not a bad face; he was not by nature a bad man."). On the other hand, Posey was also a member of "these renegade Piutes [that] had been replenished for generations by outlaws from other tribes." While the Paiutes were, according to Lyman, "fierce," "implacable," and single-mindedly intent on seeking trouble with the Mormons, the policy of the Peace Mission "was always the 'soft answer which turneth away wrath'. . . ." Yet when the two Paiute captives escaped Sheriff Oliver, "the people who had been called in 1879 to tame this wild, impetuous snarl of Piutes had the immediate task of finishing the job with their own hands. . . ." Which they did to a fare-thee-well.

The Paiutes were well aware of Mormon attitudes, and many expected that the settlers would kill them at the conclusion of Chief Posey's war. Happily, it did not come to that, although the forced division of families must have been a hard blow to an already thoroughly demoralized people.

Another Blanding acquaintance of Chief Posey, Lyman Hunter, talked with the Paiute on occasion between 1920 and 1922 before the sad ending of the Indian's life. Mr. Hunter reports, "These days, I think some people would have called Posey an ecologist. He was somewhat concerned about preservation of the land. He told me and Mancos Jim told me a time or two before how the country had been when Posey was a boy. And their expression always something like the grass would grow up to the bellies of the ponies. He said there was lots of grass and lots of deer and there was hunting."

Nowadays in southeastern Utah what little grass there is quickly fills the bellies of the cows that have the run of the range. In downtown Blanding the descendants of Chief Posey's generation drive the streets in pickup trucks identical to those owned by their far more numerous modern Mormon neighbors, the product of a small but fertile band of newcomers.

Bandidos

The Mormon migration followed an east to west route with the settlers eventually all but displacing the previous tenants of the land. I, too, am a migrant to the Southwest, but I have more in common with the Apaches than with the Mormons of my adopted state of Arizona, having gone from north to south to reach my new home, starting from the northwestern United States and retracing in a very rough way the path taken by the Na-Dene on foot long before me. I made my first trip here nonstop in a Travelall.

Kangaroo rats, not the dream of a new homeland, lured me to Arizona initially. It all started in September 1971, when my colleague Bob Lockard invited me to join an expedition he had planned to the southeastern corner of Arizona. I was then employed by the University of Washington in Seattle because Bob had somehow managed to persuade his fellow psychologists to hire a biologist, rather than one of their own, for an opening in the Department of Psychology. Luckily for me, I was that biologist and therefore was in the right place at the right time when Bob decided to head down to Arizona in a university vehicle.

Bob's goal was to locate and study banner-tailed kangaroo rats in the wild. He had picked banner-tails as his subjects after refashioning his research interests in the years just prior to my arrival at the University of Washington. Once a traditional experimental psychologist who worked with the traditional psychology department subject—the laboratory rat—Bob had become convinced that it was a mistake to rely so heavily on this almost artificial and certainly long-domesticated animal. He wrote several combatively controversial papers spelling out the deficiencies of the white rat for experimental research and the importance of a true comparative approach that would, if adopted, transform psychology into a far more biologically oriented discipline.

Bob put his money where his mouth was as he embarked on studies of wild kangaroo rats, a most unusual behavioral subject for a psychologist at the time. But Bob liked the fact that kangaroo rats were a thoroughly nondomesticated species. For example, unlike white rats, whose daily activity patterns are unaffected by moonlight at night, kangaroo rats are extremely sensitive to it. Bob already knew from laboratory work that simulated moonlight caused his captive wild-caught animals to become inactive at times when they would have continued to move around if it had been completely dark. Bob's explanation for the behavior of his "real" rodents, was that they, unlike white rats, face nocturnal predators that can see better when the moon is out, thereby favoring kangaroo rats that return to their burrows after moonrise.

Nevertheless, in order to demonstrate that the moonlight avoidance that occurred in the laboratory was not simply some bizarre artifact of captivity, Bob wanted to test whether kangaroo rats living free and natural lives were also likely to retreat to safety when the moon came up. To this end, he invented a nifty little device that could be stocked with a supply of seeds of the sort that kangaroo rats love. When a rat came visiting to collect some (and the device dispensed them slowly so that a single visit would not do the trick), it walked into the device and onto a treadle. When the treadle was depressed, it moved a pen, leaving an ink mark on a slowly rotating circle of paper that was attached to a clock-driven circle of aluminum. Each time a kangaroo rat entered Bob's machine, it made a new mark on the paper.

The point of this Rube Goldbergian apparatus was, therefore, to make a timed record of each kangaroo rat visit during an entire night. When morning came and Lockard retrieved the paper from a recorder, he could calibrate the marks with the passage of time and know precisely when the visiting rat or rats had come for a snack at his device.

Having developed this delightful invention, Bob was eager to try it out under field conditions. He somehow learned that the San Simon Valley near Portal, Arizona, was a hotbed of banner-tailed kangaroo rats. These lovely tan and white animals with their extraordinarily long, tufted tails build elaborate underground nests, creating mounds ten feet or more in diameter and two or three feet high with a maze of tunnels underneath. They are also territorial, so that when you put a timer-feeder by a particular mound, you can expect to collect an activity record just for the owner of that mound.

Bob had no difficulty convincing me and two graduate students, John Laestadius and Randy Beeton, to join him for a whirlwind tour of the San Simon Valley where we would try to secure activity records for a substantial number of territorial banner-tails. We drove without surcease from Seattle to southeastern Arizona in a big Travelall stuffed to the gills with food, traps, an immense tent, and much other miscellany. Toward the end of what seemed at the time an interminable trip, we finally reached Portal Road, long after midnight on the second day of our journey. Then and now Portal Road has a long unpaved segment, and as we pounded along the dusty corrugations, the headlights carving out a little tunnel ahead of us, it felt as if we were headed into another universe, one that was not altogether hospitable. Mercifully Bob pulled off at last to the side of the road, and we quickly set up our army cots right where we were, collapsing with thanks that we were for the moment not going to drive one more mile.

A few hours later I awoke as the sun slipped over the top of the Peloncillo Mountains, a soft low line of hills to the east of our impromptu campsite. A gathering of coyotes serenaded a quavering welcome in the cool of the morning. In the rich light of the early morning, we saw that we were in a broad valley with the Peloncillos on one side and the far higher and more rugged Chiricahua Mountains on the other. The flats around us were a mix of scruffy grassland and an even more beat-up mesquite range. And everywhere we looked there were banner-tailed kangaroo rat mounds.

Packing up our cots and sleeping bags, we drove some more until we found a dirt track leading to a stock tank a few hundred feet from the main road. Bob had been told that we would be free to camp anywhere, and the big muddy pond with its whirling windmill looked like a decent spot to us. Out came the huge canvas tent, the boxes of food and paraphernalia galore. Eventually, we wrestled the tent into place and spread our gear about the camp.

About this time we had a visit from Guy Miller. Guy Miller wanted to know what in the hell we thought we were doing putting our camp right by a stock tank. Mr. Miller proceeded to give us a crash course in camping etiquette in cow country. Rule 1, upon which he elaborated with authority, is that you do not put your camp by a stock tank. We learned that open range livestock are not fond of people. A tent could put cattle off using a tank, forcing them to go thirsty or to move a considerable distance to some other water hole. We took Mr. Miller's hint. Down came the canvas behemoth to be relocated several hundred yards away along with the rest of the expedition's ample gear.

Having become educated on rules of the range, we were soon into a marvelous week of kangaroo rat research. Some of Bob's devices had a few mechanical glitches, and one or two were stepped on by cows wandering absent-mindedly through the area. But they worked (and ultimately helped Bob show that free-living kangaroo rats often return to the safety of their burrows when the moon comes up, giving up foraging time, the better to stay alive to search for food again another evening).

The local banner-tails wasted no time learning that the patron saint of kangaroo rats had sent them a free source of millet seeds. They, unlike Guy Miller's steers, were not shy about humans or their artifacts. At night we went out with flashlights and followed kangaroo rats around as they bounced through the desert. They sometimes let us come within a few feet of them. We wondered how they could survive in a world filled with coyotes and concluded that it was lucky for kangaroo rats that coyotes do not come armed with flashlights.

The week raced by with evenings spent eating spaghetti and drinking port and cranberry juice, an eclectic but effective combination. The mornings were occupied with collecting kangaroo rat activity records, the afternoons filled with watching golden eagles and Swainson's hawks drift up and down the valley. Another rancher stopped by. He allowed as how Guy Miller sometimes got a bit hot under the collar. He also told us to keep an eye open for illegal immigrants, which often came over from Mexico just fifty miles to the south and walked up the San Simon Valley at night, hoping eventually to reach Tucson or Phoenix or some place in between where there would be work for them. He warned us that there had been a wave of petty thievery around Portal and Rodeo. One of his neighbors had gone outside to check on suspicious sounds he had been hearing around his house and had had a brick applied firmly to his head as a consequence.

Forewarned, we kept an eye open, but it was only on the very last night of our stay at our stock tank campsite that we encountered anything out of place. In the early evening, we glimpsed a couple of people at a distance in the scrub and faintly heard their conversation, so garbled by the distance that we could not be certain whether they were Spanish-speaking individuals, although they seemed unlikely to be Germans or Albanians.

I called out to them in my fractured Spanish but received no reply. The men did not leave but hovered about among the faraway mesquites. Bob became convinced that they were up to no good. Bob had a reputation for a quick temper, and he now demonstrated the validity of that reputation. He had brought with him a .38 revolver, perhaps in honor of Arizona's gunslinging tradition. Impulsively, he went into the tent to retrieve the weapon and upon returning, he quickly discharged four or five shots toward the horizon, in the general direction of but well over the heads of the distant duo.

They almost immediately disappeared, just as I would have done had I been in their shoes, and we heard no more undecipherable murmurs from them. The four of us discussed the excitement of the occasion, and then John wondered if perhaps our two mystery men might not even then be preparing to take their revenge when night came. This unsettling thought caused us to reflect silently for some time. Bob suggested that perhaps we should take it in turns to maintain a watch throughout the night.

Thus it was that at 3 A.M. I arose from my sleepless cot to relieve Randy Beeton in guarding our camp. Randy solemnly handed me a large flashlight and the immense .38 revolver that had caused all our problems in the first place before retiring to sleep the sleep of the well-protected.

For many hours now, I had not been looking forward to my role as defender of the kangaroo rat expedition. The moon offered no light on the desert whose blackness and silence at this hour provided many opportunities for introspection. As I sat glumly in camp, however, I began to realize that the nocturnal silence of the San Simon Valley was not absolute. There, just off to the right of the tent, came an almost imperceptible sound but one that to my now hypersensitive ears sounded very much like a footstep. And then another and another. Easing myself out of the camp chair, I flicked on the flashlight and directed the light beam out in the direction of the ghostly intruder. I saw nothing. I heard nothing.

I turned the flashlight off and waited again, heart rate elevated but descending. Each minute that passed (slowly) brought an increased sense of calm. The .38 felt large, bulky and foreign but if worse came to worst . . . .

In an instant, my heart was once again in my throat as a new volley of sounds ricocheted around camp. With pistol quivering in one hand, my flashlight searched wildly for the source of the noise— which proved to be a Merriam's kangaroo rat, a small relative of the banner-tails that we had come to study. The rat jumped nonchalantly out of the paper bag that it had just investigated in a malicious fit of adventure.

I now stood alert, listening to new sounds that I tried repeatedly to convince myself were merely some more small rodents disporting themselves on nocturnal expeditions near our camp. I thought that physical activity might be an antidote to the stress of the situation, and therefore I decided to walk the borders of the camp, which I did with the caution of a World War I soldier on a mission between the trenches. With my flashlight on, I inched around the Travelall when suddenly, and to my great horror, I saw a flashlight shining back at me. The heavy .38 nearly discharged of its own volition, but happily I refrained from blasting what proved to be the large side mirror on the driver's side of the Travelall, a mirror that faithfully reflected the light from my own flashlight back into my eyes.

You may not be surprised to learn that the two impoverished Mexican immigrants, if that was who they were, did not launch a murderous assault during my watch. Eventually, the first morning light came to rescue me from myself. I returned the revolver to its travel box and wearily joined my companions in arms as we began the job of breaking camp and returning to Seattle, temporarily in my case, because by the next year I was back in Arizona with a new job at a new university.

I have been back to the San Simon Valley many times but never as a free-lance camper. Poor Mexican men still make their way up the valley at night, even though the Immigration Reform and Control Act of 1986 now makes it more difficult for illegal aliens to be hired. Judging from deportation figures, which presumably comprise a tiny but constant fraction of the total illegal immigration, the flow of undocumented workers has not abated much (1676 persons were deported from Arizona in 1985 prior to the new regulations whereas the Immigration and Naturalization Service returned 1387 aliens in 1989 well after the IRCA had been fully implemented).

Guy Miller no longer has time to educate tenderfeet newcomers on the ways of the West. He has gone into the real estate business. The 3 Triangle Ranch offers 40+ acre plots with attractive financial arrangements, so they say. A competing ranch is in the hands of Empire West Real Estate; to purchase a Crown Dancer Ranch Estate you need only contact "Hoot" Gibson. Therefore I could buy, if I possessed the requisite cash, a ranchette of mesquites and gravel, still occupied by banner-tailed kangaroo rats. My mammalogist friends tell me, however, that their population has declined from the glory days when Bob, John, Randy and I came to make their acquaintance. Perhaps even the kangaroo rats are on the move these days in a world filled to overflowing with newcomers and immigrants.

Confessions of a cactus-hugger

At the peak of their power and influence in the Southwest, the Paiutes and Apaches numbered only a few thousand hunter-gatherers. Hohokam society may have been somewhat more populous, with an agriculturally supported urban system comprised of tens of thousands of people in the Hohokam heyday. However, the Hispanic and Anglo immigrants to the land of the Clovis, Hohokam and Apache have put their predecessors to shame when it comes to people production, an activity that appears to be the major modern achievement of humans worldwide.

During my roughly half-century on the planet the world's population has ballooned from a little over 2 billion to considerably more than 5 billion. According to my calculations, that means there are now an extra 3 billion people, give or take a few hundred million, compared to when I came on board in the middle of World War II. That's 3,000,000,000 plus. The worldwide addition since 1942 is the equivalent of another eleven to twelve populations equal to that of the current United States, each person consuming tons of food and producing tons of garbage in his or her lifetime, and preferring attractive living space, clean air, fresh water, a color television and a new Honda Accord or its superior.

The population eruption has not skipped the United States. Although we have not been multiplying with quite the exuberant abandon of some other nations, even so, in just the last fifty years the USA has nearly doubled its citizenry. There are so many extra people around that I personally have noticed the increase. I live in Tempe, a suburb of Phoenix, Arizona. My subdivision is now about thirty years old. When we first moved here I was twenty years younger than I am at this moment, the United States' population stood at a little over 200 million, and the nearest major street was endearingly named Rural Road. Not too long before we did our share to stimulate the already frenzied real estate market in Tempe, Rural Road passed through a largely agricultural landscape of cotton fields and sorghum plantings. Hence its name. Cottonwoods lined irrigation ditches. Aerial photographs of Tempe's agricultural fields taken in the fifties and sixties revealed faint lines, the legacy of Hohokam farmers who built irrigation ditches through what was going to become Tempe hundreds of years later, long after the Hohokam Indians abandoned their fields forever.

The ancient irrigation projects of the Hohokam are now obliterated by a zillion new housing developments, which have cannibalized all the old agricultural land in Tempe. Once you could pull out onto Rural Road at any hour of the day with barely a pause to check for oncoming traffic. Now the flood of cars running to and from south Tempe make each trip from Loyola Drive onto Rural Road a minor adventure on most days. Helicopters whir far overhead during rush "hour," offering information on which streets and freeways have become impassable and giving advice on the lesser of two evil routes home. Once you could drive out into the countryside near town and find a desert wash in which the only human footprints you were likely to encounter would be your own. No more.

I liked it the way it was. I have had it with changes. I would prefer to keep things as they are. Or were. There are plenty of people right now in Tempe, Arizona, in the United States of America, in the world. Do we really need any more? In fact, I confess that I would rejoice if the world's population were reduced (preferably painlessly) by a half.

I haven't quite reached the point of joining the Voluntary Human Extinction Movement (VHEMT, pronounced "vehement"), the brainchild of the pseudonymous "Les U. Knight" who would like to see the human species phased out altogether. But if the real estate developers of Arizona went extinct, I am sure I could adjust to their disappearance. So I am not opposed to all changes as a matter of blind principle, just the ones imposed on me by time and the press of people, the changes that rob me of the things I have come to appreciate, like desert washes that are free from beer cans, shotgun shells, ladies' underwear, cowpies, Circle K plastic bags and cups, the spoor of domesticated animals, discarded car oil, and footprints (other than my own). I rather resent my own footprints, now that I think of it. I believe that these sentiments qualify me as a genuine cactus-hugger.

My enthusiasm for cacti encourages me to head out again to the satisfyingly spiny Usery Peak, a place with more saguaros than people. As I climb up the north-facing slope of the mountain on a brilliantly sunny, unequivocally hot summer day, I have a view that is largely people-free, although I can see a few distance-diminished toy cars zipping along on faraway Usery Pass Road. The northeastern end of Apache Junction is barely visible off to my left, but to examine this classic example of suburban sprawl I have to look over my shoulder, which I rarely do since it is a good idea to keep my eyes on the trail before me. But when I come over the crest of the mountain peak and have to confront what lies spread out below to the south and west, I face an almost solid sweep of urban life. And it is a world that never stops growing. Although I was here just a few weeks ago, it has been long enough for a new epidemic of housing developments to break out on the far side of Bush Highway, a main access route into the desert near the Userys.

A new house goes up in greater Phoenix every few minutes, twenty-four hours a day, 365 days a year. It seems like just the other day that the *Arizona Republic* was full of the wonderful news that our town would exceed the magic one million mark in population for the 1990 census. To hear the newspaper talk about it, jubilation was general. I, however, managed to keep myself more or less under control.

A few days later the word came down that the preliminary

count from the census was actually 971,565. This statistic was a disappointment to those who like seven-figure numbers, among them Phoenix Mayor Paul Johnson and his companions on the City Council. On the other hand, the *Republic* was able to trumpet the good news that Phoenix had surpassed Detroit and San Antonio in population to move into the number eight slot on the list of largest American cities. (Subsequently, Phoenician politicos learned to their dismay that Detroit succeeded in turning up a sufficient number of originally uncounted persons to keep Detroit ahead of Phoenix in the numbers game.) Johnson and his fellow boosters on the City Council had even hoped against hope that we might overtake Dallas in our surge up the charts, but it was not to be, as our Texan rivals for numerical superiority were able to claim 990,957 Dallasians, or is it Dallasites?

Actually, for some years now, greater metropolitan Phoenix has topped two million souls distributed in a continuous megalopolis starting from Apache Junction in the east and running to Glendale in the west, a tedious hour's drive, if and only if the freeways are functioning during the time of travel. The growth in each of the satellite suburbs of Phoenix as well as in the city itself has been phenomenal over the past forty years. Mesa, Paradise Valley, Litchfield Park, Scottsdale, and Tempe have all exploded and thrown people right up to the borders of their neighboring municipalities.

Apache Junction is a case in point. Founded in 1950 with a population of 8,500, AJ was incorporated in 1978. By 1990 it was home to nearly 18,000 people, about 10 percent of which resided in one of the many mobile home parks that have eclipsed the long-suffering creosote flats that once occupied so much of central Arizona. The city fathers project that another 10,000 persons will be on the rolls when the year 2000 comes round. During the winter months, thousands of visitors add to the local year-round population, while perhaps enjoying a glimpse of the Superstition Mountains from their double-wide trailer home in the Ironwood Mobile Home Park or Palmas del Sol Mobile Home Park or Eldorado Mobile Home Park.

In 1924, when George Elbert Burr migrated to Phoenix, the town was not bloated with Phoenicians (of which there were less than 30,000), and Apache Junction did not even exist as a municipal entity. No one then could have envisioned the 1990 version of either place. At the time when Burr transported himself to Arizona for good, he had achieved fame as a major illustrator for magazines like Harper's, Scribner's and Frank Leslie's Weekly. In 1891 he went along with then President Benjamin Harrison on a coast-to-coast tour, generating a steady stream of prints that illustrated the expedition and helped make him even more well known as an artist-journalist.

Burr left his journalistic career behind in 1924 because of his concern for his health, which apparently was delicate for much of his adult life—although he did manage to survive until age eighty. After moving to the Southwest in search of a salubrious climate, he took to the desert in a big way and devoted his artistic talents to illustrating Arizona landscapes. His work was both prolific and a commercial success. The desert Southwest is exotic to most Americans even today, and Burr captured the strange beauty of the place in his art.

The Apache Trail provided inspiration for many of Burr's prints. The road, still unpaved in places today, begins at Apache Junction and angles up into the mountains to the east. Near its beginning, the Trail edges around the Superstition Mountains, home of the mythical Lost Dutchman's Mine. The mountains jump up out of the low plain on which modern Apache Junction and the rest of Phoenixopolis squat. The ramparts of the front range conceal wild mountains and canyons that follow one upon the other for miles toward the east.

One of Burr's more celebrated prints is entitled "Superstition Mountain, Apache Trail, Night," which he produced in 1931. The artist has us peer across a long stretch of darkened desert plain peopled with the shadowy images of saguaros, up the substantial flanks of the mountain to the palisade cliffs that make up its upper third. The cliffs are vaguely illuminated by moonlight, as is a huge cumulus cloud that hangs suspended above the peak top.

Burr's nighttime image of the Superstitions somehow manages

to convey an ambiguous sense of tranquility combined with fore-boding. The billowing cumulus clouds help create a feeling of uneasiness by hinting at the possibility of a violent thunderstorm, but perhaps the storm has already swept through the Superstitions and has begun to dissipate with nightfall. Despite the darkness of the print and the mildly ominous cloud, there is nothing gloomy about the scene it depicts. As I study Burr's mountains, I sense an acceptance of the night with its special silence and freedom from the Arizona sun.

What is most striking about the print, however, when viewed from the perspective of the 1990s, is the complete absence of any sign of humans. Sixty years ago Burr and perhaps a genuine pros-pector or two could easily have been the only persons within miles of the front wall of the Superstitions. Today an accurate rendering of the same scene at night would show a place ablaze with light from hundreds of houses, street lamps and cars, a carnival of light reaching right up onto the very flanks of the Superstitions.

The image could still be aesthetic, I suppose, but judging from Burr's desert work, in which humans rarely appear, I doubt very much that he would select the modern scene as one worthy of his time and artistic effort. The Apache Junction of 1990 has eliminated the natural night from the leading edge of what is designated as the Superstition Wilderness Area. People dominate the landscape, not the saguaros, not even the mountains themselves. There are still a few places where, if you position yourself just right during the day-time and if you have tunnel vision, you can look out on the great promontory of the Superstitions near Apache Junction and not see things human. But not after sunset, not any more now that the huge numbers of human migrants to Apache Junction have driven wildness out of the wilderness and darkness out of the desert night.

Abert's towhees and other urban opportunists

The residents of Apache Junction (and Phoenix, Mesa, Tempe, and AJ's many other urban neighbors) have replaced one environment with another, scraping off upland Sonoran Desert paloverdes, saguaros and creosote bush and replacing this vegetation with a standard suburban-urban mix of Bermuda grass and eucalyptus, processed gravel and olive trees, asphalt and African sumacs. The ecological consequences of this transformation of the land are numerous and mostly obvious, but there are also some more subtle effects as well, as demonstrated by the Abert's towhees skipping across the back patio of my home.

The uncivilized Abert's towhees of Arizona are not found in upland Sonoran desertscrub but instead have traditionally occupied a very different habitat, Arizona's streamside mesquite bosques, one of which I can easily see from the top of Usery Peak. A thin, green, double belt of riparian mesquite runs down either side of the still flowing fragment of Salt River that extends from Saguaro Lake to the Granite Reef Diversion Dam. Once upon a time a person on Usery Peak would have seen unbroken bands of mesquite lining the riverbanks for many miles more to the west, right on through modern Phoenix. But the diversion of the river for agricultural and urban uses put an end to that. Mesquites need to tap into ground water. If it is not there, they die.

When Herbert Brandt wrote *Arizona and Its Bird Life* in 1951, mesquite bosques were already well on the way out in the state, victims of development, dam building, irrigation schemes and woodcutting. Even so, Abert's towhee was not a great rarity in the remaining riparian woodlands of the fifties, although few people saw the bird because of its uncommonly secretive nature, a point that impressed Brandt.

"This trait of shyness seems to be a theme of those who have written about this retiring species. With its habitat remote from the haunts of man, why should it have developed such a timid complex, since it is doubtful that human beings have ever done it harm? However, there are few birds more elusive than this big brown sparrow, and it has always been so wherever I have been fortunate enough to find it at home."

Brandt's commentary hardly jibes with the fact that two Abert's towhees are scampering friskily across my back patio in suburban Tempe right now. They are the quintessential brown bird, pale dun brown from head to tail; only a little patch of black feathers about the beak offers any relief from their firm dedication to brownness. One bird takes five hops forward and pauses; its companion comes bounding behind and pauses. The second bird lifts and drops his long tail slightly. The other bird does the same, and then the two of them run for the leaf litter beneath the African sumac as if their lives depended on it.

Once they reach shelter of the imported city tree, they are soon hard at work, sending fallen leaves flying. Abert's towhee, like the much more familiar rufous-sided towhee, loves to kick ground litter around in the search for small insects and other morsels that hide beneath this material. After a bout of litter shifting, one bird flies up into the tree while its partner heads for the imported citrus tree nearby. The first bird gives a noisy squeal; its companion seconds the motion. Elusive they ain't.

Nowadays, Abert's towhees find the suburbs completely acceptable as a foraging and breeding grounds. Most of my friends have a pair inhabiting their backyards. When a team of ornithologists led by Ken Rosenberg censused Tempe in the mid-1980s, they calculated that in the spring and early summer there were about fifty Abert's towhees per 100 acres of suburban Tempe, a figure that made them one of the commonest bird species in town. If a non-Arizonan simply wished to add the bird to his life list quickly, he would be well advised to visit suburbia rather than hunting through a riparian canebrake or mesquite jungle, their traditional home in Brandt's time and before.

The towhee's success in making the transition to town may have a great deal to do with the determination of Arizona's suburbanites, most of them transplants from the Midwest or East, to surround themselves with greenery rather than confront the stark desert, which for much of the year is as brown as an Abert's towhee. The result has been the creation of hundreds of thousands of back-yards ornamented with imported exotic vegetation more suitable for Minnesota or New Jersey than Arizona.

As this odd habitat burgeoned in central Arizona, some pioneering Abert's towhees began to take advantage of it, probably within the last forty years. The towhees made the switch in habitat requirements just in time, because a mere 5 to 10 percent of the riparian forests that once graced our streams still remain. Those that persist are much the worse for the wear and tear imposed on them by cows and people. And when you think about it, a humid stream-side canebrake and mesquite bosque are not terribly different from a Tempe backyard bordered by shrubs and overcanopied by mulberry and citrus trees, the whole melange intermittently sprayed by sprinklers. The urban oases created by people have more than a passing, albeit superficial, resemblance to the now almost extinct native riparian forest.

Although Abert's towhees overcame their innate shyness in order to join the rowdy English sparrows, pedestrian house finches, cocky mockingbirds and pea-brained inca doves on the list of Arizona's common urban birds, many other desert birds that live outside the riparian zone have not made the adjustment. For example, rock wrens, black-tailed gnatcatchers, and black-throated sparrows are all very common upland Sonoran species that seem able to take the presence of people in stride—in their traditional habitat. However, I have yet to see a single one of these birds in my neighborhood where eucalyptus, mulberry, and olive trees have shouldered aside the creosotes, paloverdes and staghorn cacti.

My failure to spot these species does not derive from my nearsightedness. Rosenberg and his crew remarked on the absence of black-tailed gnatcatchers in their survey of Tempe, and another professional ornithologist, John T. Emlen, who conducted a similar survey in Tucson, also failed to record gnatcatchers, black-throated sparrows and ten other common desert birds in that Arizona metropolis.

It is not that city habitats are hard on birds. Both Emlen and Rosenberg's teams pointed out that Arizona's cities and towns are far more pro-avian than the surrounding desert. In Tucson, the total population of all birds was over twenty-five times as large in the city than in comparable adjacent plots of more or less undisturbed Sonoran Desert, according to Emlen's counts. But despite all the apparent advantages of the city environment, the abundant water, food, and foliage, nevertheless some desert species still turn up their beaks at the resources we so generously offer the bird world in general.

Emlen tried to make sense of the seemingly haphazard constitution of the urban bird community in Arizona. He recognized that introduced or nondesert birds constituted well over half of the total avian population of our cities. House sparrows and starlings come to us through ill-advised introductions from Europe. The extremely common inca doves have entered Arizona only in the past century, and they, too, are inveterate urbanites. Without man-altered environments they and the house sparrows disappear. The abundance in town of this coterie of birds poses no special puzzle, because these species are long-adapted to take advantage of habitats created by people.

Of the native birds that have made the transition to Arizona city-life, many are seed-eaters that consume the grass and weed seeds scattered on irrigated lawns and in urban alleys. House finches and white-winged doves fall into this category, for example, and both do well in Tempe and Tucson in the appropriate season.

On the other side of the coin, Emlen believed that ground-nesting desert birds, like Gambel's quail and rock wrens, were at a special disadvantage in town, where they rarely (if ever) occur, given the constant traffic over any urban surface. The armies of alley cats and backyard dogs also surely make life miserable for easy-to-reach ground nesters.

But selection of nesting location cannot be the whole story, because birds like black-throated sparrows and black-tailed gnat-catchers do not nest on the ground, but instead prefer shrubs and small trees in the desert. They do not avail themselves of the many equivalent sites in urban settings. Could it be that certain desert-restricted species simply lack the behavioral flexibility possessed by house finches and white-winged doves? The ecologist Jared Diamond has made precisely this argument, suggesting that some birds may be inherently more adaptable than others.

For example, species that occupy a diversity of natural habitats may be more able and willing to take advantage of novel urban environments than species that occupy only one specific habitat type. Individuals of the diverse-habitats group may have evolved a readiness to explore new situations and use resources that are different from those that they grew up with. Such an ability could help individuals take advantage of whatever exploitable habitat type they chanced upon. In contrast, members of species with more specialized and well-defined habitat requirements may be more reluctant to explore novel environments and less likely to accept the opportunities that they offer. These birds may be beautifully adapted to a single set of conditions, a habitat that has been reliably present over the millennia and in which these species flourish.

The behavioral plasticity hypothesis suggests that urban birds should be drawn primarily from the ranks of species that have demonstrated an ability to cope with variation in the natural, or at least the nonurban, world. Certainly, one of the premier urban birds of all time, the starling, has demonstrated its behavioral flexibility many times. In Europe it is widely distributed across the countryside, occupying many natural as well as unnatural habitats from northern to southern Europe. An omnivore, it can and does feed on just about everything, taking worms from lawns and insects from vegetation as well as wolfing down the seeds and fruits of many plants. Since being introduced into this country in the late nineteenth century by a misguided New Yorker, it has enjoyed spectacular success in both city and countryside, whereas many other introduced exotics have failed to maintain a toehold, let alone spread across the entire nation.

It took a little over fifty years for the first pioneering starlings to fly into Arizona, with the first sighting in 1946 and the first breeding record in 1954. The bird has multiplied with a vengeance since then and has flaunted its plasticity by moving out into the desert, where it now regularly uses the nest chambers constructed by woodpeckers in saguaro cacti. Needless to say, saguaros are a novel nesting substrate for starlings, and yet it did not take these birds long to move into the cacti and take up housekeeping in a plant that has remarkably little in common with the oaks and maples of the starling's native Europe, or its nonnative New York, for that matter.

When starlings occupy saguaros, they usurp potential nest sites for desert natives like the Gila woodpecker, gilded flicker, ash-throated flycatcher and elf owl. The competition they provide for these birds has not endeared them to ornithologists for whom the native species are infinitely more appealing than the imported interloper. Allen Phillips and his colleagues Joe Marshall and Gale Monson do not mince words when evaluating starlings in their book The Birds of Arizona.

"Their recent increase in Arizona bodes ill for our native woodpeckers and other hole-nesters such as the Purple Martin, small owls and Myiarchus flycatchers. Thus far Starlings have confined their nesting in Arizona to towns, irrigated farmlands, and adjacent saguaro desert. Perhaps they will not extend far out into the saguaros, but at any rate it is disgusting to see the Martins arriving in May to inspect saguaro holes already full of the abominable Starling families—sometimes two families in a single saguaro! These birds should have been left in Europe, where Starlings belong."

Like it or not, starlings are here to stay. They have made the adjustment many times to novel environments, including urban ones, in their man-assisted spread across the world. In sharp contrast, a large percentage of native Arizona species have not even been able to carve out a minor niche in one novel environment, the cities and towns of Arizona. Is it because they lack the "behavioral plasticity" of starlings and the other citified species? Have they become locked into habitat requirements that can only be satisfied by Sonoran Desert plant communities?

Or is it simply that they have not had the time for an adventurous behavioral variant to appear in their populations? Who would have ever guessed in the 1950s that Abert's towhees would be townies *extraordinaire* in the 1990s? Not Herbert Brandt, that's for sure. Perhaps it is just a matter of time until an innovative pair of black-tailed gnatcatchers or black-throated sparrows will establish a beachhead in Tempe or Phoenix or Glendale or Tucson. With the continuing march of urban development across the state of Arizona, whose population has nearly quintupled since Brandt's time, the sheer quantity of city habitat available for desert birds to colonize is in no danger of decreasing any time soon. One day the little gnatcatchers and desert sparrows may be as much a part of the city scene as Abert's towhees and house finches.

I am not sure whether I look forward to this day or not. It still pleases me to see an Abert's towhee prancing about on my patio, but I think I value the bird even more when I glimpse it skulking off through the Blue Point mesquite bosque, one of the few remaining patches of riparian forest near Tempe. Even here, however, city life has begun to impose itself ominously on the desert. The rumble of cars streaming across the Blue Point Bridge over the Salt River makes it hard to escape an urban frame of mind. The roar of a three-wheeler filters through the trees from where its self-absorbed owner careens cross-country through the floodplain of the Salt River, flattening everything in his self-absorbed path. Ground-nesting birds beware!

A big saguaro towers over the mesquites, many of which have had limbs severed by hit-and-run woodcutters. A starling warbles from the highest arm of the cactus, announcing his availability to some females in the neighborhood. There is a suitable nest cavity in the limb on which he perches. Is it just my imagination or is there a hint of insolence in the starling's song?

In a thicket of rough-barked mesquites, one of the remaining old-fashioned Abert's towhees pops out of the drying grasses and weeds to land on a low-slung limb of tree. Just before I manage to bring my binoculars into focus, the bird drops back down and slips away, upholding an admirable, if fading, tradition of aloofness from that ultimate urban species, human beings.

Playing God with the white-winged dove

A small flock of white-winged doves sweeps down the mountain ridge, their beautiful banded wings rowing rhythmically through the air. The color pattern of these birds is masterful, as it combines a regal gray with the slightly curved white patches that bisect the dark-tipped wings, an eye-catching arrangement.

A dove sails up to the top of a saguaro. Breaking its flight with a flurry of wing beats, it spreads its semicircular tail to reveal the brilliant dots of white that emphatically terminate each tail feather. The bird perches among a handful of fruits balanced on the trunk tip of the saguaro. The white diagonal band of its wing now outlines the lower edge of the folded wing.

Most of the saguaro fruits have burst open, revealing masses of black seeds and the blood red interior lining of the fruits. The splashes of crimson on the cactus make it look like a candidate for the accident ward. A reddish brown fluid from some overripe fruits trickles down a grooved pleat. The dove leans forward and gathers a beakful of moist pulp and seeds before sitting upright again to keep an eye on its surroundings. It tucks its head down for another mouthful and then quickly resumes a watchful air.

Eventually, the white-wing tumbles off its perch, perhaps to return with a cropful of fruit to a nestful of hungry squabs in a paloverde or ironwood. As the regurgitated mash is transferred to the youngsters, some may dribble from the squabs' beaks and fall from the nest to the ground below. There a seed or two may find a way to germinate, producing baby saguaros from the food used to help white-wings rear a family of their own.

The development of urban Arizona may conceivably have been an overall plus for Abert's towhees (if we consider only the species' total population size), but the interactions between most desert animals and their extraordinarily abundant human neighbors are complex. White-winged doves offer a good example. On the one hand, for a bird that seems a perfect part of the web of desert life, white-winged doves, like Abert's towhees and ringtail cats, have been able to profit mightily from certain activities of men. Now and in the recent past, the doves have supplemented their saguaro fruit meals with large quantities of agricultural products, a practice that has contributed to the love-hate relationship people have had with the bird. Prior to World War II, the farmers near urban Tucson and Phoenix considered the doves pests, vermin, trashbirds because they consumed seeds of sorghum, barley, safflowers, sunflowers and the like, all crops that were then grown in abundance on irrigated fields in central Arizona.

At the same time, white-wings have always been popular with hunters who, admittedly, like to kill the birds they love. Near the turn of the century, hunters gunned the doves down year-round while others collected the squabs from nests that were densely clustered in the huge mesquite bosques that lined the Santa Cruz River in Tucson. M. F. Gilman, writing in 1911, noted that "there was an attempt recently made to have them protected, but such a howl went up from the ranchers that nothing was done."

Eventually, something was done as it became apparent that a combination of unrestricted shooting and the full-scale destruction of the riparian mesquite forests through woodcutting and water diversions had greatly diminished the numbers of the doves. But even so, prior to World War II, Arizona Game and Fish opened the white-wing hunting season in early August. The birds nest into the late summer, and so many adults shot in August were breeding birds whose deaths doomed the young they were tending at the time (both parents share in incubating and feeding their offspring).

An August hunting season was not therefore in the long-term interest of either the white-wings or their hunters. L. W. Arnold believed that if it were not for World War II, white-winged doves would have been all but exterminated in Arizona. The war gave people something to do besides hunt white-wings, and after 1945, Arizona Game and Fish revised their policy so that the doves could not be shot until after the young of the year had been reared.

Without August hunting, white-wings were free to turn their attention to gleaning seeds from the vast fields of seed-producing plants that farmers tended at that time. Farming practices of the immediate post-war era included leaving the fields in stubble for up to six months after the harvest, a tactic that the doves appreciated greatly because it gave them easy access to all the unharvested seeds that lay on the ground. Their populations grew steadily, and by 1967 Arizona hunters were killing close to one million white-wings during the fall hunting season without causing a decline in the species.

But the white-wing bubble burst in the 1970s, thanks to a combination of factors. First, Arizona farmers using irrigation switched en masse from seed crops to cotton, an agricultural product that offers nothing for white-wings. Even those few that stuck with sorghum or wheat changed their farming practices so that acreage that once lay fallow for months was immediately turned over after the harvest (burying any seeds lying on the surface). The motivation for these changes lay in the development of new policies that sought to impose water conservation on farmers by requiring them to reduce the acres that they irrigated. The farmers responded by dutifully cultivating less land, but they then grew more crops per year on their reduced acreage. Their actions proved perverse from the standpoint of water conservation and dove production as well.

To make matters worse from the dove's perspective, the hordes of newcomers flocking to Arizona from Minnesota and North Dakota and elsewhere made farmland more valuable as real estate. The developers snatched up citrus groves, once an important agricultural mainstay in the Phoenix metropolitan area. Now most citrus orchards have long since been converted to rows of houses. When the real estate developers bulldozed the acres of citrus trees, leaving just a few in some of the upscale developments as landscape ornaments, they unknowingly took out prime nesting habitat for white-wings, a replacement breeding grounds for the mesquites that an earlier generation had destroyed. The white-winged dove, like the passenger pigeon, often nested in amazingly dense colonies in mesquite groves and citrus farms with up to 250 nesting pairs an acre in the good old days. No colonies of this sort can form in cities that offer only scattered trees and lawns of Bermuda grass.

As if the disappearance of citrus groves wasn't bad enough (for the doves), the city of Phoenix had the riverbed and banks of the Gila River cleaned up and channeled after a series of damaging floods in 1978, 1979, and 1980. In the course of this work, they destroyed a great stretch of salt cedar. This tree was introduced into the United States from Europe or the Middle East in the late 1800s and has since spread its feathery branches along most of the watercourses of the Southwest. In some places, salt cedar forms nearly impenetrable thickets that smother native vegetation and block the free flow of water down riverbeds, thus the effort to rid the Gila River of the weed. As a rule, aficionados of native Arizonan plants shed few tears for deceased exotic imports, but in this case, the salt cedar jungle on the Gila, like some citrus groves, served as a vast colonial rookery for white-winged doves. When the salt cedar went down under the bulldozers, so did the doves that depended on it for nesting habitat.

The outcome of the ups and downs affecting white-wings has been that the annual "harvest" of doves by hunters has now fallen to one-tenth of what it was during the heyday of the bird. It is hard to know how to feel about the decline, because in some sense, the post-war population explosion of the bird in urban areas was completely artificial, depending as it did on agricultural sources of food and exotic nesting habitat of citrus and salt cedar groves that would never have existed without our heavy-handed intervention in the affairs of the desert. On the other hand, the citrus groves and salt cedar stands merely replaced what an earlier generation had taken from the birds, namely the mesquite bosques and riverbank vegetation that had supported colonial white-wings in the past.

Although the great colonies of the bird are largely a thing of the past, white-wings fortunately do not require the close company of hundreds of their fellows in order to reproduce. Unlike the passenger pigeon, which *always* nested in vast hordes and apparently could not revise its reproductive requirements when its population fell sharply as a result of over-hunting and forest clearing, white-wings will nest on their own as scattered pairs both in urban settings as well as in native riparian and desert environments. Thus, through no fault of our own, the bird is hanging on reasonably well. Despite its population decline, it is still one of the standard urban birds in Arizona. But who knows what changes we have in store for our environment in the future, changes that may send the population of the dove on another roller-coaster ride. For better or worse, the dove's destiny, like that of salt cedar, citrus trees and Abert's towhees, is now in our hands, the same hands that more skillfully guide bulldozers on their appointed rounds.

DESERT CATTLE

A cow is a very good animal in the field;
but we turn her out of a garden.

SAMUEL JOHNSON

Cows

The little stream in the Chiricahua Mountains has mined its way into ground too stony for kangaroo rats, forming a steep-sided ditch that weaves down the mountainside. In midsummer, only a trickle of water slips over the rocks at the bottom of the trench. The roots of tough-leaved oaks and ponderosa pines try to hang onto what little soil remains on the slopes above the water. A painted redstart works its way up the trunk of an oak, repeatedly spreading and closing its white outer-tail feathers like a card shark flourishing a hand of cards. Beneath the brilliant warbler, the floor of the forest is covered by the dull detritus of dead oak leaves and pine needles.

Down by the water, things are somewhat more lively, thanks to a border of bright green grasses. Casual inspection of the grasses, however, reveals that something has mowed the narrow green margin nearly down to its roots. Even the willows trying to grow by the water's edge have been attacked, bent, and fractured, and are missing most of their leaves. There is little doubt as to the culprit, which is not a ground sloth, mammoth or gomphothere, not even a deer, one of the larger of the North American mammals to survive the megafaunal crash of 11,000 years ago. In the water, a fat lump of cow dung circles sedately in a little eddy. Other cowpies, dried and blackened, decorate the shore, and hoofprints of cattle cover other hoofprints there. A little pool in the stream has begun to fill with black silt eroded from the stream bank. This is cow country in the Coronado National Forest. Here the principle of multiple use of federal forest lands is put into action every day that cows exercise their grazing rights on the allotment that includes the middle fork of Cave Creek.

The Apaches are long gone from the Chiricahuas and in their place are Anglo ranchers, who have amplified their presence many times over by bringing cattle with them into the mountains. After the disappearance of the ground sloths and glyptodonts, there were no hefty grazers in the Southwest for ten thousand years until the reintroduction of the horse and the novel addition of cows in the mid-nineteenth century. Our cattle are extensions of our modern selves. While we cluster in cities, steers are hard at it in the country-side, acting as agents of ecological change in every patch of desert and isolated mountain canyon.

From up the hillside among the oaks comes the herniated bel-low of a cow separated from her calf. More than a hint of her ribs ripples her black and white washboard hide. Her pelvis juts out in anything but a jaunty fashion. A vacant look has long since made itself at home in her eyes.

Unlike the distant Userys, where grazing leases have been retired for a number of years, here in southeastern Arizona cattle still are at home on the range in most places in the public domain. The Chiri-cahua Mountains are far more representative than the Userys when it comes to livestock. Almost everywhere in Arizona you will find cows, some fat, some skinny, some tame, some skittish, all hun-gry. In this state, the United States Forest Service is responsible for more than 11.8 million acres of allotments to ranchers, including Mr. Guy Miller's Cave Creek lease. And the Bureau of Land Manage-ment, known informally to some as the Bureau of Livestock and Mining, administers an additional 12 million plus acres of grazing leases in the state.

One might imagine that all the livestock feasting on such a hand-some amount of real estate would constitute a really impressive bovine army. In turn it could support a battalion of ranchers and contribute formidably to the state's economy, thereby accounting for the political power wielded by livestock interests in Arizona. In reality, just 150,000 head of cattle graze over these millions of acres, a figure that computes to one cow per 152 acres of public lands. This is not prime cattle-raising territory.

"I swear to God, you absolutely have to despise your cows to put them out there," claims Arizona's land commissioner, Jean Hassell, speaking of the desert range that state officials lease to ranchers. Which is probably why all the cows produced on range-land throughout Arizona and all the rest of the West add up to just two or three percent of the nation's total, the overwhelming majority of which grow fat on feed lots eating corn in Illinois and Iowa and other midwestern locales.

Arizona cows have to do the best they can on the generally arid and largely grass-free country that the State Land Commission and federal agencies manage for Arizona's ranchers. Only a handful of these ranchers, just 1500 or so, have secured grazing leases on all these millions of acres. Among the 1500, a few operate on spectacularly large chunks of Arizona. For example, John A. Whitney, Jr., and John A. Whitney III have the permit for 294 square miles of the Tonto National Forest on which they are allowed to run 1,174 steers. The Crowder-Weisser Ranch near Quartzite has the BLM's permission to nurture 1,400 head of cattle on 351 square miles. Throughout the West, less than 3 percent of the owners of grazing permits control nearly 50 percent of the grazing land administered by the BLM. The remaining 97 percent of those who have grazing permits in the West are therefore reduced to presiding over relatively small spreads. One study claimed that the small and large operators collectively generate $300,000,000, by generous estimate, for the state's economy. (This figure was manufactured by multiplying by three the sum of $100,000,000 actually produced by rural cattlemen on the grounds that each dollar of "cattle" money passes through three hands in rural Arizona.) If we accept the livestock industry's figure of $300,000,000 generated toward the cash economy of the state, some of which comes from the grazing of public lands, the next question is, how much is $300,000,000? The answer is surprisingly little compared to other sources of revenue for the state. For example, in 1990 tourists more or less happily deposited $6,000,000,000 into the willing hands of Arizona merchants and motel managers, about twenty times as much revenue as is said to be produced by the livestock industry.

Or take the Motorola company. This one manufacturer had a 1990 payroll that was more than twice as large as the $300,000,000 that rural cattlemen claim as their contribution to Arizona's economy. Motorola spends more money on goods within Arizona in a year than all the cattle business in the state is worth.

Or take one week's worth of retail sales in greater Phoenix, which comes to $320,000,000, according to the folks who calculate these things.

Still one could argue, and some have, that cattle-based income may be a small contribution overall but to the good folks out in Safford and Kingman and Patagonia, it still means a great deal. To continue the defense of cows, one could argue, and some have, that with only 150,000 cows spread out over more than 20 million acres of rangeland, it's hard to see how they can do that much damage. Moreover, it is possible that steers are merely ecological replacements for the mammoths and ground sloths that once grazed this land.

But in the era when there were no cows and no people around, the mammoths and ground sloths had to contend with an impressive array of predators that are sadly no longer with us. These predators almost certainly would have helped keep a cap on the population of their prey and would have kept them on the move as well. In contrast, today's cattle have the countryside pretty much to themselves. They are free to gather in numbers and remain for months on end in those parts of the landscape where the grazing is best. Not surprisingly, cows love a pleasant stream where there is water to drink and an abundance of food of which they need a great deal if they are to survive and reproduce. A rangeland cow moves 1000 pounds of forage into her mouth each month. A lesser but still substantial amount exits from her anus in the same period. In an arid land, the processing of this much vegetation per cow can easily make an impression on the local environment, particularly in the narrow bands of streamside vegetation where desert cattle aggregate. It is no accident that the stream itself and immediate environs bear the brunt of cows on the Cave Creek Grazing Allotment.

The understandable fondness for water and forage that cattle exhibit can create havoc on the range during times of drought, times that are not uncommon in Arizona. For example, the customary and much appreciated summer rains all but failed to materialize in 1989 around Phoenix. Because about half the annual rainfall comes (as a rule) from thunderstorms in July, August and early September, their absence in 1989 meant hard times ahead for desert life. Three inches less rain in a year may not mean much to an Easterner, but three inches is a major shortfall in an area where seven or eight inches is the whole enchilada.

By fall the native trees and shrubs, which normally flourish at this time of cooler temperatures and higher soil moisture levels, had come to suffer in silence. Out in Randolph Canyon in the Superstition Wilderness Area, about 30 miles from the equally dry Usery Mountains, the mesquites were all but leafless, as were the brittle-bush and bursage. Nothing at all bothered to flower. The desert vegetation seemed to have all but evaporated, like the water that was once in the soil, leaving behind just the outlines of plants.

But cows, and plenty of them, still searched the canyon relentlessly for a bite to eat. Even though the stream that sometimes flows in the canyon had dried along 99 percent of its length, water still persisted in a set of deep depressions cut by past floods into the bedrock of the stream. The water kept the cows alive, after a fashion. They survived to devastate the terrain up and down the stream around Red Tanks, hunting for forage in a land that simply did not have much left. The nearest vaguely green things were far up the steeper hillsides. A person could walk for hundreds of yards around Red Tanks and never be more than a few feet from a cowpie, most as dry as the dust in which they lay. One cow also lay in the dust, on her side, her belly full at last but with methane, not desert vegetation. Bloated, dead, and stinking, she had collapsed in a thicket of equally dead gray weeds. Her brilliant fluorescent yellow ear tag was the only lively thing left about her.

Range management is not my forte, but even I could discern, I thought, a pattern of overgrazing in Randolph Canyon. In my letter to Mr. James Kimball, a forest supervisor in the employ of the U.S. Forest Service, I wondered why the USFS permitted so many cows to remain on drought-stricken desert lands. Mr. Kimball replied courteously and at length, acknowledging that there was a problem on the Millsite Allotment but claiming that the grazing permitee, a Mr. Martin, had recently agreed to a new system in which the area would receive a twelve-month rest from grazing after eighteen months of sustaining a herd of cows. Mr. Kimball cautioned me that the improvement in forage conditions would be gradual and that "past overgrazing practices are not going to be corrected overnight." After having seen what Mr. Martin's cows had accomplished during their stay in Randolph Canyon, I had no doubt that Mr. Kimball was on target. I found it hard to take solace from the fact that Mr. Martin's cows were doing their small bit to contribute to the rural economy of the state of Arizona.

One man's campaign

In Arizona and over much of the rest of the west, rural tradition has it that a cow is to be treated with the kind of respect that would not be considered excessive by adherents of Hinduism. A reflection of this attitude is that here cows have the right of the road, a point that may impress you or your heirs after you run into an Arizona steer on a night when you are doing seventy and the steer is stuck in neutral on the white dividing line. Should you join the steer in another world, the steer's owner may be able to add insult to injury by collecting damages from your next of kin to compensate for the loss of his animal.

But the law, like the environment, can change. In December 1990 the Arizona Supreme Court ordered a modification of the policy on cow-car encounters, despite the time-honored legal practices of our state. Now, if you can prove negligence on the part of the cow's owner, you (or in your absence, your kin) can counter-sue him for damages. That's the good news. The bad news is that "negligence" does not mean just letting the steer get out of a fenced field to wander bewilderedly onto a highway. Unless you can demonstrate that the stockman shooed the beast into the path of your onrushing vehicle, it is going to be tough to collect in court. But it is a kind of moral victory to know that there are at least some conditions under which the legal status of cows does not surpass that of cow-consumers in Arizona.

Cattle still rule supreme when it comes to cattle-feed. Consider what happens when a gang of delinquent cows decides that the grass is greener on your side of the fence. When they manage to push your fence down or step over it or through it to reach your private property, which they then treat as their private grazing reserve and defecatorium, you have only one option, which is to calmly chase the malefactors off your land and repair your fence. You have no legal right to claim recompense for your losses from the cows' owner. And you would be well-advised not to injure one of the intruders as you seek to reclaim your real estate.

Vince Roth knows all about the legal ramifications of the treatment accorded uninvited cows on one's private property. As a long-time director of the Southwestern Research Station in the Chiricahua Mountains, Vince had many opportunities to observe cows in action on the Forest Service lands adjacent to the station. He didn't like what he saw as the cows roamed the range on neighboring hillsides, removing forage with great efficiency. Once they had eaten the better part of the grass on their grazing leases, the local steers often cast an envious eye onto the station grounds, which were nibbled only by the native deer and so always had a comparatively luxuriant carpet of grasses. The temptation was so great that some of the more enterprising (or hungrier) steers managed to find a way through or around the station's fences.

Vince did not enjoy chasing after his unwanted guests and ushering them back across the dividing line between the Station and the National Forest. Then one summer day something snapped in Vince when he saw yet another trespasser on his turf. This steer had proved to be a veritable bovine Houdini, constantly finding new ways onto the station's lawns where Vince would find the beast grazing energetically but contentedly, secure in the knowledge that nothing other than a mild frolic about the lawn awaited it as punishment for its transgression. But on this dramatic day, Vince went into his house and returned with his shotgun, which he discharged at the rump of the gate-crashing steer. Shortly thereafter the animal headed back on the double to the thoroughly grazed public lands that were its legitimate, if somewhat the worse for wear, feeding grounds. The birdshot that peppered the beast's behind was intended to impress upon the steer the need to respect Vince's views on cattle grazing and private property. I can report that the birdshot appeared to have exactly this gratifying effect.

But the shotgun blast also had some unexpected repercussions, because up on the National Forest hillside above the station, one of the Mexican cowboys in the hire of the grazing lessee happened to observe Vince in the very act of educating the errant steer. What the cowboy saw evidently also impressed him, because he informed his boss of the day's melodrama, which was when the cow flop hit the fan, to paraphrase an old Southwestern saying.

The local ranchers had long known that Vince harbored a certain distaste for cows and the grazing practices of cow owners. Their resentment of Vince now blossomed into outrage as the ranchers saw a chance to remove a thorn from their collective side. Threats of legal action and retaliation made their way up to the Southwestern Research Station, and a letter-writing campaign calling for Vince's removal as director reached the American Museum of Natural History, the patron of the station and employer of Vince. In addition, Vince came in for punishment of a special sort from the locals. He was and is a long-time devotee of the country-swing dances that occur on Saturday nights in the bars and halls of flea-bitten hamlets like Rodeo and Animas. Now he was told that he and his crew of biological visitors were not welcome at these occasions anymore. The word went out among the ranching community that anyone socializing with Vince ran the risk of being ostracized.

Vince took the loss of dancing privileges hard, but happily for him the sheriff of Cochise County never came to take him away nor did the director of the American Museum heed the advice of the Chiricahuan stockmen who called for his forced resignation. And to this day, he still casts a jaundiced eye on the sacred cows of the

West and recalls with some satisfaction the day when he took the Law in his own hands at high noon in Cave Creek Canyon.

Vince Roth was in his own way a pioneer, exploring new methods to deal with the great grass consumers of Arizona. In more recent years, the radical environmental group Earth First! has taken up where Vince left off. They are believed to have escalated attacks on cows and the means to keep them eating what is available on public lands. Or so the papers report, with occasional stories on stock killing and vandalism of windmills by unknown persons, actions that get attributed to Earth Firsters! because some prominent members of the group have advocated extreme measures to make grazing on public lands unprofitable.

According to the radical anti-cow contingent, if a rancher were to lose even a few steers on the marginal lands his cattle exploit, his profit margin would evaporate. Or if his windmills were sealed with concrete, they would no longer pump the water needed to keep his cows fat and happy in desert or semi-desert terrain. Cut a cattleman's fences, and his cattle will wander off, perhaps to be lost forever or at least to require considerable effort and expense before they are recovered.

These kinds of activities are, needless to say, illegal, and I suspect that Earth Firsters! are mostly talk and little action. But even the talk has made Arizona ranchers nervous and encouraged them to believe some interesting rumors. It must have been a rancher who passed on information for an editorial in the *Mesa Tribune*, which warned darkly of a new tactic of dedicated eco-terrorists, the castration of prize bulls of ranchers whose cattle occupy public lands. The image of a bearded Earth Firster! creeping cautiously toward a rangeland bull armed with a loop of wire is an intriguing one even if it has precious little correspondence with reality.

My own guess is that the number of really dedicated eco-terrorists is depressingly small. Therefore, a war against ranching on public lands that depends on Earth First! or originals like Vince Roth is bound to be a prolonged affair and one that is not likely to have a clean resolution. But these "extremists" may have a small role to play in making others who seek to change the grazing situation through more conventional means appear more respectable in the eyes of the general public. If nothing else, their actual or imagined activities add a certain frontier flavor to conservation battles round and about the West.

One of the more conventional ways to try to effect social change is through legislation, and as I write, Congressman Mike Synar, a member of the House of Representatives from Oklahoma, is giving it a go. Representative Synar would seem to be an unlikely advocate for change in the ranching industry. He was twice a 4-H vice-president for the State of Oklahoma, is now a rancher running a cow-calf operation, and in the future he promises never ever to become a vegetarian. But despite his background, Mr. Synar volunteers that "a few federal grazing permit holders are feeding off the Federal Treasury," a proclamation that has made him about as popular with western livestock interests as a timber wolf in a corral.

But Representative Synar does not shoot from the hip. He comes armed with numbers, noting that there are 1.6 million cattle producers in the country, exactly 2 percent of which graze their steers on BLM or Forest Service land. Even out west, only 8 percent of the cattle growers are public grass consumers. It is a marvel that so few can have so much to say about public policy and receive such a handsome amount of federal money. And the handouts to the minority are indeed substantial. Mr. Synar notes that the fee for grazing a cow on BLM or Forest Service lands in 1990 was $1.81 per month. It is true that this figure inched up to $1.97 in 1991, but the fee has a long way to go before it matches what the Department of Agriculture calculates as the commercial value for grazing one cow on public lands for one month, which they place at $8.70.

Representative Synar believes in a "pay as you go" basis, and he has put legislation where his mouth is, with a bill that would gradually phase out the current grazing subsidy. If he ever succeeds in getting a sufficient number of congressmen to agree with him, the average taxpayer and the conventional cactus hugger will have some reason to celebrate. Those small cow-calf operations that depend on the subsidy for their profits may have to call it quits. The large corporate enterprises will at least have to pay a bit more for the privilege of running cows over the western landscape.

Critics of the Synar approach point out that chasing a few marginal ranchers off the land will do little to reduce the number of cows run on public lands for the simple reason that current federal law *requires* that grazing allotments be grazed. Failure to do so exposes the permit holder to the risk of losing his allotment to someone willing and able to put cows on the "unused" acres. The sensible thing to do, so this argument goes, would be to change the law that makes grazing mandatory on most of the public lands in the West and, at the same time, to institute rigorous controls on cattle operating in sensitive riparian areas.

These critics are on target as far as they go, but unhappily, Congress does not appear to be on the threshold of doing the sensible thing. Even Representative Synar's admittedly incomplete reform has failed on several occasions to become the law of the land. But the bill would be one small step for desert conservation and, judging from the cries of anguish coming from the cattle industry, or at least the tiny part that uses public rangeland, the increased fees would make public lands grazing a less attractive commercial venture. Perhaps some abandoned allotments would find no cattleman eager to run cows on BLM or USFS terrain at his own expense, now that taxpayers were not available to pick up the tab. Perhaps Congress would be more willing to retire grazing leases from land the cattle industry no longer wanted. I can hope, can't I?

Cattle free in 1893

Below the Chiricahua Mountains where Vince Roth fired the shot that was heard around the San Simon Valley, the land flattens itself, washes peter out in fingers of sand, and mesquites take control of the terrain. These are not the fat-trunked, patrician mesquites with over-arching canopies that line up shoulder to shoulder along permanent streams to drink the water there. These are spindly, thirsty, plebeian mesquites, many of which are less than six feet high. They possess not one central trunk, but an array of skinny rods that poke up through the banner-tail kangaroo rat mounds and support a ragtag collection of leaves. The spiny outer twigs of the dwarf mesquites shake in the breezes. Pale brown sparrows huddle in their shade.

Out in the open a pair of black dung beetles urgently pushes a ball of cow dung across the nearly grassless plain, as they search for a place to bury their drab prize. Once underground, the big marble of dung will receive an egg from the female, the egg will hatch into a grub, the grub will feast on the buried dung and eventually become an adult dung beetle in search of fresh cow dung to produce a new generation of beetles. The newly adult dung beetle will not have far to search.

Cows by the hundreds graze in the San Simon Valley, shaping the mesquite "trees" as if they were bonsai masters. They keep the plants crouching close to the land where they can be clipped back again and again whenever grasses are especially scarce and the cows are particularly hungry.

Although now grasses are not abundant in the valley even in the best of seasons, once upon a time Cave Creek steadily flowed far out into a broad plain between the mountains that was more prairie than scrub chaparral, where grama grasses flourished rather than tattered mesquites. The anti-Apache author Samuel Cozzens wrote that in the 1860s the "valley of the San Cimon is about twenty-five miles in width, and contains much fine grazing land, as well as some good agricultural districts. It is covered with a species of grass called *grama*, which for its nutritious qualities is rivalled only by the celebrated mesquit [sic] grass of Texas."

Cows have been eating grama grass and contributing to cash flows in Arizona for less than 200 years, but some people believe that this has been long enough to utterly transform the once extensive grasslands of the territory. Throughout southern Arizona

as recently as the late nineteenth century, rich grasslands dominated the terrain in places that now have the look of desert about
them. The replacement of grama grasses with mesquite trees co-
incided with the disappearance of permanent streams, which are
now dry washes down which water rushes only after a summer
storm drenches the watershed. The surging waters persist for a
short time, but at the height of their power they cut fiercely into the
stream bed, creating steep banks and scenes of desolation in places
that long ago were pleasantly pastoral. What caused these unhappy
changes?

In trying to solve the mystery associated with the disappearance of grasslands and the dramatically altered nature of streams in
southern Arizona, many persons have fingered cattle as the prime
suspects. In the late nineteenth century, after the campaigns against
the native Indians in the Southwest, notably the Apaches, had
achieved their goal of forcing these inhabitants into exile, a few
entrepreneurs began to stock the now Apache-free range with cattle. Some pioneer ranchers raked in the cash during the early years.
As the word of their success spread, it fueled an immense cattle
boom with the numbers of cows calling the Arizona range "home"
rocketing from 5000 in 1870 to 35,000 a decade later, and over
650,000 by 1883. In February 1885, the *Southwestern Stockmen* examined whether Arizona's grasslands were overstocked but dismissed
the possibility as "remote." Today there are only a fraction of the
number of head roaming the range in the state as there were in
1885, and even this much reduced herd suffices to produce some
spectacular cases of what intense grazing can accomplish.

The nineteenth century buildup, however, was not finished by
1885. More Texas steers poured into the state, boosting the cattle
population over the million mark by 1890 and up to perhaps as
many as a million and a half in 1891, a year when the monsoon
rains failed in southern Arizona.

The years 1892 and 1893 brought with them a flat-out drought
during which hundreds of thousands of cows died, so many that
one observer reckoned that you could make your way across

Arizona and always be within a stone's throw of a dead steer. Before dying, the doomed cattle naturally ate everything in the vegetable category within reach. By the end of 1893 the editors and subscribers of the *Southwestern Stockmen* must have realized that heavy grazing and drought had created apocalyptic conditions in Arizona's grasslands. Now this was about the time that grasslands began their metamorphosis into mesquite chaparral and permanent streams began their conversion into deeply cut dry washes in Arizona. The correlation suggests that the orgy of overstocking and consequent overgrazing caused the two changes in the landscape. Such an explanation has obvious plausibility given that cows unquestionably consume grasses, clearing the ground for pioneering mesquites as well as removing plant cover that would retard the movement of rainfall from the land to the streams. Increased run-off could create surges of floodwater, which definitely do restructure the streams in which they occur, cutting deep channels in places where broad, shallow, permanent flows once meandered.

But although the cow-as-culprit hypothesis has many adherents, fairness requires that we consider an alternative explanation, namely the possibility that climatic changes destroyed Arizona's prairies and permanent streams. If the late nineteenth century saw a shift to drier years coupled with a more compressed summer season of violent thunderstorms, then the prairies may have succumbed not so much to cows as to a shortage of rainfall. The streams may have changed their character because they did not have sufficient rains to maintain year-round flows, and when it did rain, it did so explosively, causing destructive flash flooding.

In their instructive book *The Changing Mile*, James Hastings and Raymond Turner examine several ways in which to test the climate-change hypothesis in lieu of complete and accurate meteorological records for the period, which, needless to say, do not exist. For one thing, they point out that in the early- to mid-nineteenth century, there was another cattle boom (and bust). Between 1820 and 1846 (when the range was abandoned due to Apache attacks), Mexican ranchers ran as many as 150,000 head in southern Arizona. At this time the territory was claimed by Mexico and occupied by Mexican settlers. Concerted Apache assaults, not drought, put an end to this first episode of cattle ranching on a grand scale in the area. But while the ranchers held their ground, their huge herds seem not to have caused vast environmental disturbances because the prairies persisted until the 1880s and 1890s.

Still, although great numbers of cattle roamed the plains of Arizona in the 1840s, perhaps ten times as many chewed their way through Arizona later in the century, making comparisons on the impact of cattle in the two eras suspect. Happily, there is yet another way to gather evidence on the relative importance of climate change versus cattle grazing on the ecology of the desert Southwest, and that is to examine the history of a place where cattle have never placed their heavy hooves on the land. Admittedly, finding such a place in the Southwest is extremely difficult. Even today in the most remote areas and on the steepest, driest hillsides you are likely to find either cows or the sure evidence of their presence.

But Hastings and Turner did succeed in locating a continuously cattle-free zone and, even better, one that had a photographic record extending back to 1907! This remarkable spot is MacDougal Crater in the Pinacate Mountains of northwestern Sonora, Mexico. The Pinacates are among the most remote and isolated of regions in the world, and for good reason. The annual rainfall is on the order of five inches, and in many years even less falls. The landscape is covered with fresh lava flows so much like a moonscape that the Pinacates attracted American astronauts as a training site for the first moon landing. Despite the apparent impossibility of productive ranching in such a spot, Mexican cattlemen have, in fact, for many years pushed some cows into the Pinacates to gather what they might from the region.

Within the Pinacates, however, there exist great craters, vast sunken amphitheaters with sheer rock walls that offer no easy access for people, let alone cows. Volcanic activity produced the craters, either through the explosive force of superheated ground water or through the formation and then collapse of underground chambers. Either way, the results have been impressive with eleven major craters in the area, some with diameters of almost a mile.

Although Amerindians doubtless knew of these craters thousands of years ago, they have been discovered and rediscovered by European and North American explorer-adventurers several times in the past 500 years. One expedition, led by Daniel MacDougal and William Hornaday, entered the area in 1907 and made some photographs on their trip of what was later named MacDougal Crater. These black-and-white prints have been well preserved and offer a clear look at what vegetation grew in the Crater's cattle-free floor in 1907. Hastings and Turner have been back nine times, beginning in 1959 when they located the exact vantage point from which certain photographs were taken by Daniel MacDougal a half-century earlier. The modern sequence of photographs by Hastings and Turner, coupled with on-site censuses of desert plants in specific plots on the crater's floor, offers a chance to compare past and present, and to identify what has happened over an eighty-year span.

The changes in vegetation within the crater over this time have been surprising and revealing. For example, the numbers of saguaro cacti increased from 38 in the 1907 photograph to 159 in the 1977 duplicate. In the same period, the number of creosote bushes fell by a half—from 103 to 53—and declines of similar severity occurred in the population of paloverdes. But even as certain shrubs and trees were becoming rarer and rarer, mesquites were doing spectacularly well. In one plot where just two mesquites grew in 1959, a mini-forest of 186 individuals had formed by 1982.

The results of Hastings and Turner's work on the floor of Mac-Dougal Crater provide convincing evidence that desert plant communities can change dramatically over a fairly short time, especially with respect to the abundance of mesquites—without assistance from cattle. The changes in vegetation in the Crater are unlikely to have been caused by anything other than climatic events. The key phenomena, according to Raymond Turner, are probably episodic bouts of severe drought, which are known to have occurred within the Pinacates or nearby regions at various periods in the past century.

Droughts can kill. Although creosote bush looks like it could survive forever without water, this species and paloverdes are apparently more drought-sensitive than saguaros and mesquites. By 1959, a severe drought had been going on for more than a decade, during which time the numbers of creosotes and paloverdes plummeted. Then in the early 1970s, two tropical storms roared in off the Pacific and surged through the Pinacates. The abundant rain provided by these rare autumn storms created perfect conditions for seed germination. But for reasons that are not entirely clear, only mesquites were able to take advantage of the opportunity (perhaps because their seeds are particularly long-lived).

If similar climatic vagaries affected much of southern Arizona in the 1880s, then there, too, livestock may have been irrelevant as causes of environmental change. Given the MacDougal Crater study, we cannot rule out the possibility that mesquite would have spread and streams stopped flowing across the landscape of the Southwest at the end of the nineteenth and early part of the twentieth century even if the cattle boom and bust of the 1880s had never taken place.

I confess that there is within me some element that finds the MacDougal Crater study inconvenient, an annoyance. Part of me, presumably the cactus-hugging part, wishes that we could point with unfettered certainty to cows as the certain primal agent of environmental degradation, past and present, in the Arizonan landscape. Nor is this substantial component of my psyche much mollified by the final conclusion of Hastings and Turner that perhaps climate *and* cows played interactive roles in the desertification of Arizona and the spread of scrub mesquites across the state at the turn of the century. They suggest, reasonably enough, that overgrazing could have amplified the damaging effects of decreased rainfall on desert prairies.

But the views of Hastings and Turner have not won universal acceptance. Other persons, notably Conrad Bahre, argue forcefully that stream downcutting and mesquite invasion are the product not of climate change, but human intervention. According to Bahre, one critical aspect of Anglo settlement may have been the prevention of the prairie fires that once were a regular feature of southern Arizona. Not only did Anglo newcomers work directly to suppress wildfire, they received ample indirect assistance from their bovine companions. The huge numbers of Arizonan cattle in the late nineteenth century removed the fuel base for prairie fires, which were needed to keep trees from invading grassland terrain. Frequent fires keep mesquites burned to the ground, while permitting fast-growing grasses to dominate the scene.

If Bahre is correct, we can in good conscience hold cattle and people primarily responsible for a vast reworking of the Arizona landscape. But Bahre is unlikely to have had the last word on this contentious matter. Debate will almost surely continue on the possible explanations for the grassland and stream alterations that took place in the late nineteenth century. No one ever said that it would be easy to understand the basis for an environmental transformation that began about a century ago.

The difficulty of the task can be measured by the inability of people to agree about the *contemporary* effects of livestock grazing. This is true despite our access to reasonably accurate data on climatic variables, livestock populations and range conditions for the past several decades. Although many are inclined to assign negative environmental effects to livestock, an equally impassioned group favors the argument that grazing is a benign force on the land. Among the pro-grazing advocates are three university professors, Jeffrey Mosley, Lamar Smith and Phil Ogden (a fellow Arizonan). These academics assign the title "myth" to the assertion that livestock grazing has degraded public lands (and they go on to identify six other myths related to the cattle industry in their booklet on cows and grazing).

The three cow-huggers, all of whom are range management specialists, play their trump card early by noting, correctly, that "public lands are in the best condition that they have been in this century." A skeptic might, however, comment that at the beginning of the century in question public lands in Arizona (and elsewhere in the West) had been raped, pillaged, trampled, and devastated by the most extreme grazing pressure this country has ever seen. Thus, the baseline selected by Mosley, Smith and Ogden makes it remarkably easy to claim that things are better now than they have been at their worst, which was very bad indeed.

Much of the range managers' essay devoted to the "myth" of environmental deterioration focuses on the ratings of range condition used by the Bureau of Land Management and the Forest Service, ratings that run from excellent to very poor in four or five steps. These scores are supposed to represent the extent to which the actual vegetation at a site resembles that which would be there were it not for some human-related use, primarily grazing by livestock. By the BLM and Forest Service's own accounting, which is probably more forgiving than that provided by, say, the Sierra Club, over half the lands they manage deserve a "less-than-good" rating.

Not surprisingly, a gaggle of environmentalists have taken the low scores to mean that over half the public lands open to grazing are currently being trashed by cows. This conclusion has distressed the BLM, the Forest Service, and the publishing combine of Mosley, Smith, and Ogden. The authors of *Seven Popular Myths* argue that merely because the actual vegetation on a site differs greatly from the plants that would grow there in the absence of livestock does not mean that the site produces inadequate amounts of "desired plant species." I think I am safe in guessing that what they mean by "desired plant species" are those that ranchers desire, speaking on behalf of their cows, or those that range managers desire, speaking on behalf of ranchers. Clearly, if the "desired" plants were ones that would grow there only if cattle were removed, it *would* be appropriate to conclude that rangeland rated as "fair" or "poor" really was in moderately to absolutely miserable condition.

Because most conservationists interested in public land condition take the BLM and Forest Service ratings at face value, much to the discomfort of these institutions, the feds have come up with an ingenious solution to the problem. They intend to replace the traditional terms for rangeland condition, terms like "good," "fair" and "poor," with new ones, namely "late-seral," "mid-seral" and "early-seral." Perhaps the new jargon came from the same committee that invented "aerial interdiction of the enemy" to replace "bombing" and "collateral damage" to mean "dead civilians." The promulgation of the new code might be taken by a cynic to mean that the BLM and Forest Service actually do have something to hide, possibly the environmental damage done by the cows that they have sponsored on the public lands in their care. I, for one, still think it is entirely possible that today's livestock, although less numerous than the population that flourished briefly in the West during the late nineteenth century, are still doing a number on "potential natural communities" or PNCs as they are now called by Forest Service personnel in the know.

The impact of an impact statement

The south fork of Cave Creek ducks and bobs through the Coronado National Forest, which here consists of conifers and oaks squeezed into a narrow canyon bottom trapped by parallel cliffs of orange-red rhyolite. Within the corridor of greenery, the creek angles first one way and then another. The water in the stream drops into a pool and climbs out again to scamper down a riffle. Then it disappears entirely underground only to re-emerge triumphantly a couple of hundred feet downstream.

A black-tailed rattlesnake slides sluggishly from one resting place beside the trail to another beneath a lichen-covered boulder. The faint murmuring of the creek and wind rustling among the pines merge to create a tuneless muttering on a sleepy summer afternoon.

Suddenly, a big green and crimson bird interrupts the somnolent atmosphere with a sharp flurry of wingbeats as it sails in from nowhere to land on a bare limb on a dead pine. The bird is as gaudy as a Christmas ornament and ten times as large. It pauses briefly with a big crumpled dragonfly gripped firmly in its pincer-pliers beak before flying a short distance farther to a cavity in a pine. As it lands, a large nestling pokes its gray head up to the nest opening and receives the dragonfly from its father, which dives off among the oaks immediately after the transfer of prey has occurred.

The several young birds in the nest pipe together for a second or two after the departure of their father, and then quiet returns to the woodland again. The creek spatters between orange and white stones, adding a syncopated beat to its customary white noise symphony.

The nesting bird belongs to a species called the elegant trogon, a species that you cannot find in the Userys or the Superstitions. In fact, it is not a particularly abundant bird even in Mexico, where the bulk of its population occurs. In the United States a total of perhaps fifty pairs breed most years in a handful of mountain canyons in the southeasternmost part of Arizona and New Mexico.

The south fork of Cave Creek is one of the most reliable places to find elegant trogons in the United States. Often a pair nested right at what used to be a small, primitive campground at a trailhead, where bird-watchers gathered to admire the bird and add the species to their life-lists. The elegant trogon is every bit as elegant and admirable as its name suggests. The male in particular, with his crimson belly and green back and coppery tail, has shortened the breath and elevated the heart rate of many a birder who tracked one down for the first time in South Fork Canyon.

Many of these same bird-watchers have been surprised and even dismayed to come across cows and their calling cards in the course of wandering along the stream in search of elegant trogons and painted redstarts. Although for most administrators in the U.S. Forest Service, cows are a valuable manmade addition to the fauna of the Southwest, they leave much to be desired according to those in bird-watching and cactus-hugging circles. Some people in these groups contacted the Forest Service back in the mid-1980s as the USFS began to develop a new management plan for the Chiricahua Mountains. One option for the managers was to make part of Cave Creek a Zoological-Botanical Area (a ZBA). The proposed ZBA contained a portion of the breeding habitat of the elegant trogon, which is merely one component of a distinctive flora and fauna found in Cave Creek Canyon and a few neighboring sites but nowhere else in the United States. Many of those favoring a South Fork Canyon ZBA felt that the unique biological properties of the region deserved the special protection that would come from retiring the grazing lease from a relatively small area, just 762 acres.

The Forest Service seriously considered the idea and included it as a possibility in their draft management plan. They invited and received letters galore filled with comment on the various alternatives for dealing with Cave Creek. I was reminded that I was one of the letter writers when I came across a copy of my handwritten letter, composed on July 25, 1985, in a magisterial two-volume set entitled *Public Comments and Forest Service Response to the DEIS, Proposed Coronado National Forest Plan*. The tomes contained every letter written by members of the public with their views on one part or another of the Draft Environmental Impact Statement (DEIS) on the Proposed Land and Resource Management Plan for the Coronado National Forest. My letter dealt primarily with the various options listed in the DEIS having to do with the south fork of Cave Creek and under what conditions it might become a ZBA.

I was surprised to see my letter not only because I had forgotten that I had written it, but also because of the response it elicited from the Forest Service, which I was able to read because it appeared neatly typed on the page opposite my letter in Volume 1. (Perhaps I had received this response in my personal mail, but if so, I had forgotten this event, too.) There it was, laid out in three sections in reply to three points that the Forest Service discerned within my letter. Their answer was far longer than my letter, and much of it dealt with a single sentence that I made in closing, in which I proclaimed that "the decision to permit grazing to continue [in the proposed South Fork ZBA] sickens me. I know the ranching interests put a great deal of pressure on the Forest Service, but just once it would be great to tell them that there are other things in life besides subsidized cattle feed."

The Forest Service really took off in their rebuttal of my views on welfare ranching, and for a time I thought I personally must have touched a raw nerve at headquarters. But later, as I went through the compendium of letters and replies, I realized that anyone who mentioned subsidized ranching, whether briefly or at length, received the same detailed rebuttal. Sensibly enough, the Forest Service developed a series of standard replies to issues that came up over and over again in the various letters, and one of these issues was the question of whether ranchers receive undue assistance from the USFS.

Here's some of what an anonymous Forest Service employee had to say about the subsidy issue.

"As a matter of principle, we do not believe that it is appropriate to single out Forest Service range permitees as being the sole recipients of a federal 'subsidy.' Practically speaking, a subsidy exists whenever an individual receives benefits in excess of the fees paid by that individual to enjoy those benefits. . . . For example, in the case of a recreation user who pays no fees, or only a nominal fee, to use the National Forest, the user is receiving a considerable benefit without having to pay the full actual cost of providing that benefit. . . . This kind of subsidy would include virtually all developed recreation use, dispersed recreation use, skiers, bird watchers, wilderness users, hunters and fishermen, . . . anyone who uses a Forest Service road, etc."

I had to confess that whoever he or she was, he or she had made a number of telling points in his or her small lecture on the question of who gets what from the feds. Many times I have seen the unfortunate Forest Service employee on his garbage pick-up mission, cruising along in a pale green Forest Service truck heading up or down Cave Creek Canyon in the Chiricahuas on the way from one Forest Service campground to the next. I have camped at these campgrounds, and I have deposited trash in the redolent trash cans there myself. I have seen new signs appear at trailheads, guiding recreational users like me to the often wonderful trails that the Forest Service maintains after a fashion within their domain. I have indeed driven my car over Forest Service roads and have probably admired wildlife that have drunk their fill at a stock tank installed on public lands by a rancher with Forest Service assistance.

So perhaps I better shut up about this subsidy business if we are all in it together. But then again, maybe not. First, before I get combative, let me express thanks to the Forest Service for that portion of their budget that is devoted to bird-watchers, campers and others of their ilk. It's the least I can do and I accept the fact that people like me are to some extent subsidized by Washington. In 1988, the cost of the recreation program of the USFS was nearly $90,000,000.

On the other hand, without getting too bitter about it, I do think that there is a distinction to be made between the very small number of grazing permitees and the very large number of tourists, hikers, campers, photographers and the like who use the National Forests. Each grazing permitee receives a bundle in subsidies (totalling about $50,000,000 in 1990) compared to each recreational user. That's distinction number one.

Second, the rancher's use of public lands is extractive, designed to take something out of the forest and convert it into cold cash for the rancher's personal benefit. A grazing lease is worth a lot of money for this reason. For example, John Whitney's lease to run 1,700 cows on about 188,000 acres of central Arizona is valued by Valley National Bank at $800 per head, allowing Mr. Whitney to borrow against the roughly 1.5 million dollar value of his permit. If you or I were to purchase Mr. Whitney's ranch and allied public lands grazing permit, you or I would pay him (not our representatives in the United States government) the full dollar value of his permit. Would-be ranchers are willing to pay up, suggesting that they expect to get their money back, and then some, once they get their cows on the range.

I'll grant you that hikers and campers also do a certain amount of environmental damage as they tramp about, eroding trails, filling up the garbage bins and outhouses at Forest Service campsites, and messing with Indian ruins. But these public-lands visitors aren't using up the land to turn a profit.

However, even if I had the time or inclination to reply in full to the Forest Service's response to me back in 1985 (and who knows, maybe I did), I have the feeling that the ranchers would still have carried the day. People who can afford to spend hundreds of thousands of dollars for a grazing permit generally pack plenty of political wallop.

As it was, the south fork of Cave Creek *was* declared a ZBA but on the Forest Service's terms, with only a small area so designated and cattle grazing sanctified as part of a "carefully-regulated management plan." The cows that are allowed in with the trogons are permitted to make off with "only" 30 percent of the edible vegetation in the ZBA. As the Forest Service pointed out, cattle grazing has been going on in the South Fork Canyon for years. I know that it's a tradition there and elsewhere, but I cling stubbornly to my belief that it's a tradition that doesn't deserve to be subsidized anymore. The impact of my views remains to be detected at Forest Service headquarters.

Mountain lion mathematics: A report from Klondyke, Arizona

The impact of cows, whether on the south fork or the north fork of Cave Creek, whether in woodlands or grasslands, whether in Forest Service or Bureau of Land Management terrain, involves more than just their direct effects on desert streams and desert grasses. They also kill mountain lions.

I have never seen a mountain lion, and I regret it. I can make the admittedly far more modest claim, one that is not likely to inspire envy, of having seen a mountain lion scat once in the mountains near Burro Creek in western Arizona. There it was, lying close to a mound of burro dung on a ridge littered with loose rocks and Apache tears, the smooth black marbles of basalt cast out by an ancient volcano. I felt confident that I was admiring a mountain lion's scat because its form so closely resembled the dried, twisted feces of cats that I removed in years gone by from my boys' sandbox when they were children and more recently from my garden, a favored latrine for the neighborhood cats. By comparison, the mountain lion scat was gargantuan, daunting in its dimensions. I felt some relief that my garden is not visited by pumas.

The ridge ornamented with the lion scat overlooked a maze of narrow canyons labelled "Hells Half Acre" on the U. S. Geological Survey Map of the area. The pale yellow canyon walls, deep cut channels, and jumbled mountains there create the kind of rugged wilderness well suited for mountain lions, which are well-camouflaged, secretive and not comfortable around people.

Despite the retiring nature of mountain lions, Eddie Lackner has seen a lot of them, generally during the last moments of their lives. According to several reports in the *Arizona Republic*, Mr. Lackner runs his cows on land administered by the United States Forest Service in and around the Galiuro Mountains, near Klondyke, Arizona. He, his wife and his sons have secured grazing leases for more than 14,000 acres of Coronado National Forest, on which they are permitted to run 135 cattle the year round. That's one cow per 100 acres, which tells you something about the quality of grazing provided by this land. But it is good mountain lion country, and Eddie Lackner has the legal right to kill any predator that harms any of his cattle, whether the predator operates on his private property or on the public lands for which Mr. Lackner possesses a grazing lease.

In Arizona today only mountain lions and black bear pose much of a threat to cattle. Dick Miller killed the last Arizona grizzly bear in Stray Horse Canyon near Clifton, Arizona, in 1935. He said that if he had known it was a grizzly, he would not have gone after it. The last Arizona wolf appears to have been shot in 1960. A bounty of $100 was paid for a dead mountain lion as late as 1968 in Arizona. During the years in which the bounty system was in force, hunters cashed in 5400 lions. Despite these losses, as of 1990 about 2500 lions (and 3000 black bears) still existed in Arizona.

In November 1987, a deer hunter in the Galiuros happened upon a black bear that had been caught in a huge grizzly trap. The bear had been left to die a slow death. The hunter found several more dead bears in the area, which was, as it turns out, part of Eddie Lackner's Squaw Basin grazing allotment from the Forest Service. The hunter notified Arizona Game and Fish, and officers from this agency discovered still more carcasses of trapped bears when they went out to investigate in Squaw Basin.

Now, although ranchers can freely kill predators that endanger their livestock, they are required by law to report their successes to Arizona Game and Fish in a timely fashion and to check their traps once a day. Giant grizzly traps are illegal to all but ranchers because they endanger people who might inadvertently stumble upon one. Stand a grizzly trap on its end and it will reach up to your waist. Lackner had not only (legally) used monster traps, he had (illegally) made no effort to report his bear kills, nor had he bothered to check his traps daily.

The local wildlife manager employed by Arizona Game and Fish filed several charges against Lackner on these grounds, and the news of Lackner's offenses made their way into the newspapers, causing a certain amount of indignation within various groups of bear-huggers. But not at the Arizona Game and Fish Commission nor at the U.S. Forest Service nor at the state legislature.

The Arizona Game and Fish Commission ignored the charges brought by the local wildlife manager responsible for the Klondyke region. They imposed no civil penalties on Mr. Lackner, although a criminal case against him did result in two years of probation and a suspended $277 fine.

The Forest Service also withdrew (for two years) one of his two grazing leases, the Four Mile allotment, but not Lackner's Squaw Basin lease, which is where he set his grizzly traps. When asked why the Forest Service had devised this odd penalty, Range Staff Officer Larry Allen replied that they were going to take this action even before the bear brouhaha because the range had been heavily grazed and was in need of a rest from Lackner's cows. Lackner, who manages the Four Mile allotment for his sons, simply moved sixty-five head of cattle from this area to other public lands that he leases from the state of Arizona.

The legislators, for their part, responded to pressure to revise the antiquated anti-predator laws of Arizona by developing a new bill that permits a rancher to dispatch one large predator for each cow that the rancher claims was lost to a stock-killer. One calf, one mountain lion. One steer, one black bear. The rancher must notify Arizona Game and Fish of his actions, and preemptive strikes are not legal; but as critics point out, these regulations are not substantially different from those already in place. Moreover, the enforcement capacity of the state agencies has not changed one iota. Should a rancher forget to notify Arizona Game and Fish of a mountain lion he has shot, he is unlikely to be found out, and, judging from the Lackner case, even if he is, he will incur no significant penalty.

There is, however, one major difference in the new regulations, and that is that the names of the ranchers who legally kill big predators on public lands will no longer be public information. Arizona's ranchers lobbied hard for the right to remain anonymous in these cases because they feared they would be targeted for retaliatory action by Arizona's heretofore all but invisible coterie of eco-terrorists. Some Arizona ranchers actually contribute to their problems with predators by allowing their bulls to breed year round with the result that their cows drop calves in all months, enabling the local cow-hunting lions or bears to pick the youngsters off one by one when they are small and vulnerable. Some persons have suggested that ranchers on public lands should manage their herds so that all the cows will give birth at the same time in an area where the calves can be protected until they are large enough to take care of themselves. Others have argued that these ranchers should also be willing to accept some losses to lions or bears as part of the bargain that comes with leasing Forest Service or BLM property. At the time he had his difficulties with the law, Lackner was paying the government less than $23.00 per year for each mature cow that he ran on Forest Service lands. It was costing the Forest Service at least three times that much per cow to administer Eddie Lackner's leases.

A mature cow sells for somewhere between $500 to $1000 these days in Arizona. Mr. Lackner's entire herd that feeds on public lands is therefore probably worth in the neighborhood of $100,000. Like most Arizona ranchers who use public grazing lands, Mr. Lackner runs a cow-calf operation in which the young animals about a year old are sold off to feedlots where they put on most of their weight before slaughter. Calves and steers are, of course, worth much less than a mature cow, so that Mr. Lackner's annual sales of beef raised on his 14,000-acre allotment are surely worth far less than $100,000. To help Eddie Lackner and some of his fellow ranchers each grow, say, $20,000 or $30,000 worth of calves a year, a very large federal bureaucracy has been constructed at substantial public expense.

Another part of the tab that the taxpayer picks up for Arizona's ranchers is the bill from the Animal Damage Control division. This USDA agency employs about a dozen hunters and trappers in Arizona who hunt and trap predators and other vermin at the behest of ranchers and farmers. The ADC does its work throughout the United States but primarily in the West, receiving federal funding to the tune of nearly $30,000,000 in 1988. In return, the ADC generates an annual body count of about 4.6 million enemies of the farmer and rancher (based on records from 1988). Although most of those that are sacrificed for the public good are blackbirds of various sorts, the ADC also zeros in on coyotes and mountain lions with 76,000 coyotes and 203 mountain lions biting the dust in ADC operations in 1988.

In 1989 and 1990, the ADC helped Mr. Lackner raise calves by destroying more than three dozen mountain lions on or very near Lackner's ranch and leases in the Coronado National Forest. There is no report on how many more went to their reward courtesy of Eddie Lackner himself. The land Mr. Lackner rents from the government evidently provides good habitat for lions, and the area under his stewardship is large, about 45 square miles. However, even in superb mountain lion terrain in Arizona, one female needs and defends about 20 square miles to sustain herself and her kits. Males are less fiercely territorial, but John Phelps, an Arizona Game and Fish biologist, believes that, at most, five adult lions might be unlucky enough to call Eddie Lackner's ranch their home. Barry Burkhardt, outdoor editor for the *Arizona Republic*, reports that Arizona Game and Fish officials have compared five with three dozen plus and concluded that they better check out the situation down in Klondyke. "That's a bunch of lions, chief," says Game and Fish employee John Phelps.

In fact, it is more mountain lions than exist in all of Florida, where there are perhaps as few as thirty panthers, as the mountain lion is generally called back East. In Florida, the state legislature designated the mountain lion to be the state mammal in 1982. They did so at a time when the population of the panther had fallen to about fifty individuals. The honor of being state mammal (and a listed endangered subspecies since 1967) has not done a great deal for the panther, judging from its continuing decline.

In recent years, a combination of Floridian and federal agencies have collectively spent a million dollars annually in the effort to save the Florida panther from utter oblivion. About half the surviving members of the Florida subspecies carry radio collars permitting the panther recovery team to track their every movement, a necessity given the perils that the panthers face there, but a sad necessity nonetheless. The same fate has yet to be imposed on Arizona's mountain lions, but when our legislature proclaims the mountain lion to be the official state mammal of Arizona, you and I won't be able to find a mountain lion scat in Hells Half Acre or anywhere else, no matter how hard we try.

More mountain lion mathematics

The Arizona Cattlemen's Association seems a little nervous these days. There has been a lot of talk recently about Cattle Free in 1993, and so the ACA came up with the cash for a study conducted by George Seperich of Arizona State University's School of Agribusiness and Environmental Resources. In the course of his research, Professor Seperich discovered that, among other things, Arizona cattle ranchers annually perform 47,930 transactions in local banks. To promote this and the other findings of the ASU study, as well as to deal with some additional public relations matters, the association has produced a small brochure, which appears to be printed on recycled paper.

One of the topics discussed in the pamphlet is "the conflict with predators." In four succinct paragraphs, the cattlemen's spokesperson points out that there is an "ever-growing population of predators" ready, able and willing to eat ranchers out of house and home. Although the ACA brochure claims that ranchers "value mountain lions, bears, coyotes and other game," their enthusiasm for these species is tempered by concern for the wildlife that mountain lions and their kind consume. Prey species "suffer greatly when there are too many predators," according to the ACA handout.

Oddly, despite the cattlemen's perception of a population explosion in big predators, Arizona Game and Fish estimates that the total number of mountain lions in the state has fluctuated between 2200 and 2500 over the past ten years. We are in no danger of being overrun any time soon by mountain lions.

From the mountain lions' perspective, there has been an alarming population explosion in Arizona. In the last decade alone, another 750,000 humans have been added to the rolls in the state. Although Arizona is still one of the least populated states in the nation, for every extant Arizona mountain lion, there are now about 1500 extant Arizonans.

Although mountain lions are vastly outnumbered, the Cattlemen's Association has been highly impressed by their fondness for beef. The bottom line is still the bottom line, and ranchers have, as you might imagine, been keeping track of their losses to mountain lions, bears and coyotes. In the six years between 1983 and 1988, these losses have been calculated (by the ranchers themselves) at $4 million with mountain lions supposedly responsible for about $1 million of this amount. However, elsewhere in their brochure, the cattlemen mention with some pride that they sell $500 million worth of cattle each year, a figure that puts the losses attributed to predators in some perspective. On an annual basis, Arizona's mountain lions have been able to inflict a loss amounting to less than one-twentieth of 1 percent of the sales generated by Arizonan cattlemen, a bit of mathematics that appears nowhere in the ACA report.

Cowpies

There are some mountain lions left in the Superstition Mountains, far enough away from the Animal Damage Control agency to breathe easy for the time being. I won't see one today, but even in the absence of monsoon storms, which have failed to materialize this July, the "actual natural communities" of the Superstition Mountains still have much to recommend them: scattered tufts of golden grasses gleaming among the saguaros, black-throated sparrows flitting from teddy-bear cholla to teddy-bear cholla, a line of cottonwoods all shaking their leaves together in response to an erratic breeze. But this new drought has once again all but eliminated the stream in Randolph Canyon. A tiny trickle of water seeps from a cracked rock wall at the edge of the stream. Dozens of yellow paper wasps cover the thin fingers of water oozing down the superheated wall. They drink as deeply as they can before carrying fluid back to colony mates guarding paper nests hung beneath rock ledges nearby.

Farther along, a surviving remnant of stream flows an inch deep and a yard wide for fifty feet before submerging beneath the deep loose gravel of the dry streambed. Mr. Martin's cows peer at me from hiding places in the shade of streamside cottonwoods and willows. A host of leopard frogs bound back into the shaded litter away from the thin trickle of water as I approach. Cowpies litter the trampled approaches to the streamlet, which supports a heavy growth of filamentous algae.

A cowpie is not the most aesthetic object in the world, particularly when placed in or near water in an official wilderness. But "unaesthetic" does not necessarily mean "useless." Humans have actually found a surprising number of uses for humble cowpies. No doubt the most familiar of these is the employment of dung as a fertilizer. There are, however, many other, more exotic, applications. Cow-chip hurling contests have occasionally entertained a select company of westerners. On a more practical plane, thoroughly dried cow dung is also handy as a fuel. Untold numbers of campfires and cooking fires have been kept aflame with desiccated cow chips in places where woody matter is in short supply. And in Brazil hundreds of thousands of rural homes have been sealed with a cow dung and sand mortar, eliminating the cracks that house a disease-transmitting reduviid bug, or "assassin bug" in English, "barbeiro" in Portuguese. The bugs are sometimes infected with a species of trypanosome. When the barbeiro pierces the skin of a sleeping person to consume his blood, it may pass the parasite on to its victim, insuring either early death or a lifetime of medical misery for the unlucky human.

The formula for crack-proof cement was provided gratis by a South American bird, the rufous ovenbird or "hornero" (in Spanish), which builds a large, two-chambered mud nest, whose appearance is reminiscent of an outdoor oven of the sort once common in rural Argentina and Brazil. The mud is composed of cow dung and sand. Once it has dried, it is odor-free and extremely durable, a matter of importance to nesting horneros, which build their nests right out in the open where any passing predator can see them— but not break them apart to retrieve the eggs or nestlings within.

Thus it is not just humans that find utility in cow dung. Horneros do not occur in the American Southwest, but we do possess plenty of dung beetles as well as another aficionado extraordinaire of cow dung in our abundant desert termites. Most people have not seen a single southwestern dung-eating termite because these insects are both small and subterranean. Despite their ability to avoid attention or comment, the role of termites in the economy of the desert is remarkably large.

This point has been established by one of the few termite-watching teams in the world, a group led by Walter Whitford, then at New Mexico State University. Working at sites in New Mexico, he and his colleagues attempted to measure the impact of termites on the breakdown of organic materials in various ways. In one experiment, they collected forty-eight cowpats, freely donated by New Mexican steers. They then dried the pats thoroughly before placing them out in a 12 × 4 grid, with each pie 1.5 meters apart. One-half of the cow chips were placed on soil that had been treated with chlordane, an insecticidal termite killer, while the other half were carefully deposited on untreated soil.

By sampling the cowpies at intervals over four months, Whitford and his fellow termite counters established that termites love cowpies. The average cowpat on untreated soil had 273 termites in it when sampled after the summer rains had begun. Through their collective efforts, termites had removed anywhere from one-fifth to all of the material in a given cowpat by the time the study ended. In contrast, the cowpies that stood on insecticide-treated soil were, as expected, termite free, and they were also nearly untouched by other consumers when the research team ended their work. Whitford estimates that sans termites, a New Mexican cow-pat would require twenty years before it decomposed completely. There seems little doubt that without the recycling efforts of dung-eating termites, the grasslands of New Mexico would in short order become an unbroken carpet of cowpies, which would smother the vegetation, much to the detriment of the ranching industry, among other entities.

Termites not only remove and break down dung, but they feast on dead grasses to such an extent that they may equal grazing cattle in terms of grass consumption in the desert. Those persons who calculate what the desert can stand in terms of cattle numbers do not factor in termite grazing, leading to an overoptimistic estimate of how much forage is available for cattle on our public lands.

Incidentally, in the process of transporting cattle dung and desert grass into their underground burrows, termites also move a tremendous amount of dirt to the surface. William L. Nutting and his co-workers at the University of Arizona measured this aspect of termite ecology by clearing a plot of Sonoran Desert land of all plant material. They then placed several hundred rolls of toilet paper on the ground in a neatly organized grid. The area's termites of necessity accepted this only remaining source of cellulose, into and over which they burrowed, building their enclosed mud tunnels up from their underground base. When the termite team gathered up samples of the rolls they had distributed, they were able to shake out and weigh the dirt in the termites' protective tunnels. From these measurements, they calculated that the termites in one acre of desert bring about 650 pounds of soil to the surface each year.

Therefore, despite the insignificant size of any one termite, by working together in vast colonies that may contain up to 300,000 members, desert termites are major shakers and movers in desert ecosystems. In cow country, they receive a considerable part of their energy supply from the dung that cows generate in such abundance.

Although the biologists in New Mexico did not find that any other animal was much interested in the energy contained in dried cow chips, Vernon Bostick suggests that in some parts of the Southwest there is another animal that can make good use of the calories and nutrients that remain in cowpies. The animal that Bostick has in mind is the desert tortoise. Writing in *Rangelands*, a journal of range management, Bostick notes that the great and alarming decline of desert tortoise populations occurred *after* passage of the Taylor Grazing Act in 1934, an act that resulted in a reduction of 50 percent in the number of cattle on western rangelands.

Bostick thinks that the two events are causally linked, with the tortoise in trouble precisely because cows are much less abundant than they were in the past in the Mojave and Sonoran deserts. His argument goes as follows. The tortoise digests green plant food very slowly. On certain diets of fresh plant material, the beast can actually lose weight, even when given as much food as it can consume. But, according to Bostick, the digestive efficiency of the tortoise would be much improved if it could consume some "pre-digested" food. Calling the feces of cows "pre-digested food" somehow reminds me of the used-car salesmen's use of the term "pre-owned" to describe second-hand cars. But Bostick claims that the tortoises' nutritional "stress could be relieved if [they] had access to their natural food source, cow dung." It is the "shortage" of cow dung, according to this hypothesis, that has led to a plummeting desert tortoise population.

It is a bold hypothesis and one that flies right in the face of the conventional wisdom that cattle compete directly with the tortoise for plant matter. The argument as presented by Bostick does make one wonder what desert tortoises did before there were cows around to provide them with their "natural food source," but perhaps the reptiles once made do with the dung of deer and peccary, and in the more remote past, the droppings of glyptodonts, ground sloths and mammoths.

If Bostick's hypothesis were true, we would expect that observers of the tortoise would have seen the animal eating cow dung, of which there is still a great deal in most tortoise habitats, albeit perhaps not as much as in the 1930s. However, Vernon Bostick did not offer direct evidence on dung-feeding by tortoises, perhaps because he had none. I have not seen tortoises dining on cow dung, but then again I rarely find the animals on my walks, and when I do, the reaction of the reptile is usually to stop whatever it is doing and withdraw into its shell where it waits with greater patience than I possess until I go on about my business.

Once I did have the good fortune to see an unaware tortoise snipping off bits of the green leaves of a low-lying desert annual.

And on another occasion, I happened upon a feeding turtle that had its back directly to me. While I remained frozen in place, the tortoise turned slowly and ambled over to a patch of dried desert grasses. It failed to see me because it was blind in its right eye, the one that was responsible for detecting potential predators in the place where I stood. Not knowing that I was less than ten feet away, the tortoise proceeded to stretch out its long and wrinkled neck to reach a mouthful of yellow grass stems. Its beak closed on the stems, and the tortoise cut them off with a pull and twist of its head. A pause and another dry mouthful, then another, as it crunched its way through a most unsucculent meal of sun-dried stems.

To learn what others have seen tortoises consuming, I went through substantial lists detailing the plant species that appear in the diets of tortoises, check sheets that include buckwheats, many grasses, prickly pear cactus, and globe mallow, to name some of the edibles that the animals have been seen ingesting in their slow-witted fashion. Cow dung is conspicuous by its absence from these lists.

Thus, I am skeptical of Bostick's thesis, although I grudgingly admire the chutzpah of his attack on tortoise-hugging, anti-cow conservationists. It may be bull, but it's bull dung on a grand scale, a big enough bullpie to sustain a whole colony of 300,000 desert termites.

Peccaries

Half-way through my day-long hike in the Superstition Mountains, I realize that I am more than half-way on the road to developing several blisters, not just one, but a complete set. This surprises and distresses me because my feet have not blistered in many years and my boots are thoroughly broken in, so much so that they are coming apart at the seams, which might be the cause of my trouble. Or is it my socks? I consider the alternative hypotheses and contemplate the miles that lie ahead while I have lunch.

I have stopped by a small, nearly dry wash, which shoots down a hillside and across the trail on its way to a more significant wash far below. On the other side of the canyon the slope ascends to a distant ridgeline where rock spires line up side by side in military array to march along the horizon.

My perch is a smooth boulder overlooking the little wash, which has carved a baroque channel into solid rock. A black-chinned sparrow ducks into a shrub near the channel and stays just long enough to be identified. As I work my way through a peanut butter and jelly sandwich, I am joined by a canyon wren on a luncheon mission of its own. The wren moves between a pair of deep potholes gouged into the rock below my boulder bench. The bottoms of the potholes are littered with bits of fallen rock and gravel, and each has a small green patch of barely moist scum, the remnants of the pools that once occupied the depressions. The organic debris has attracted a gathering of little flies, and the chestnut and white wren scampers about pouncing on these insects, slipping like a mouse through tunnels formed by rocks lying on rocks in pursuit of its next morsel.

The wren proves to be a master fly-catcher. It simply snatches up the flies too sluggish to try to escape and, like a duelist armed with rapier, lunges forward in short dashes after more agile victims, its long beak snapping and slashing. Occasionally, it leaps into the air with a flurry of wingbeats after flies that have become airborne. The wren is always on the move, hurrying, hurrying as if its ultimate mission in life is to catch as many flies as possible. If this is its goal, it seems well on the way to realizing it.

After my less frenetic lunch I leave the canyon wren to its nonstop exertions as I resume mine. Hobbling along the trail, I feel nowhere near as lively as the dedicated wren. The trail undulates up and down canyons, cresting a rise only to descend by stages and then to regain the lost elevation later.

Part-way down one desiccated hillside, a company of javelina or white-collared peccaries startles from their resting place. Two of the pigs run off snorting in confusion, ultimately traveling in a semicircle, perhaps because they don't know quite where I am. As a result, one winds up only twenty feet away, peering thickheadedly about with the hairs on its back erected in fear and its nose twitching in an effort to pick up my scent. When I move again, the two sprint into the tangled desert chaparral. This time, they follow a straight trajectory and disappear among the corrugations of the hillside. An occasional clatter of loose rocks keeps me informed for a time about their progress across the slope.

Down the trail, five more javelina burst from the cover of a thicket of brown-leaved jojobas and barrel downhill in various directions with syncopated woofs of alarm. A pod of three keep together. They slow down only after they cross the dry wash at the base of the canyon. Afterward they trot more and more sedately up the hillside on the other side of the wash, slowing, then stopping to blend into their bone-dry surroundings.

It is always a thrill to see peccaries in the desert because I encounter them only occasionally. Still, javelina accent my hikes more often than coyotes or mule deer, and they seem an utterly natural part of desert life, thoroughly capable of carving out a livelihood even at times when drought has erased most traces of green from the canyons.

But the fortunes of white-collared peccaries, like those of desert termites and desert tortoises, are thoroughly entangled with the activities of cows, which may be responsible for the recent expansion of peccary range into Arizona. Most of the early explorers that traveled in Arizona through what is now fine peccary habitat did not include the animal in their lists of mammals seen (and shot). There are a few exceptions, indicating that the pig was present in the mid-1800s but probably in much lower densities than currently. Since the turn of the century, peccaries have clearly become more abundant in southern Arizona, and in the past forty to fifty years, they have invaded mountain ranges in central and western Arizona where they were entirely absent not so long ago.

Especially strong evidence in favor of the newcomer hypothesis is that the Indians of southern Arizona, the Pima and Papago, do not have a word in their own languages for the peccary but instead have borrowed one from the Spanish. As keen observers and consumers of nature, these peoples would almost surely have devised their own name for the highly edible peccary had they been present in even small numbers prior to the nineteenth century.

The population explosion in peccaries, assuming that one occurred, appears to have taken place about 1900, coinciding with the spread of mesquite across what were once Arizona's grasslands. Mesquites produce an abundance of mesquite beans, which the pigs appreciate and consume. Moreover, degraded grasslands are prime habitat for prickly pear cacti, the creme de la creme of peccary chow. Therefore, it is a reasonable idea that the changed Arizonan environment resulted in new food supplies for peccaries, which took advantage of these resources to become far more numerous in Arizona than they had ever been before.

To the extent that cattle contributed to the creation of prime peccary habitat, the current abundance of peccaries is a by-product of the cow invasion of Arizona and consequent heavy grazing, the flip side of the bovine suppression of native grass populations. I am little surprised that the Arizona Cattlemen's Association has let this one slip by, because they have been aggressive in educating the public on the virtues of cattle grazing, not just in terms of the economic returns for the state but also because of supposed environmental "benefits" as well. The cattle lobby has gone so far as to speak of the beneficent effects of cattle hoofprints, which break through the crusted soil surface and, according to the cattlemen, permit water to penetrate the desert earth more easily.

I limp up the trail, gingerly creating my own set of faint footprints, which I fear are not good for anything. I pass prickly pear in abundance and an old corral, weathered into aesthetic grays and browns, miles from the nearest extant road. Tufts of grass poke up next to the pitted rocks that lie scattered on the hillside. From the distant cliff on the right, a canyon wren lets loose its wonderfully liquid call, a descending scale that whistles down from the lichen-covered rock wall. Provided canyons persist in Arizona, we can expect to have canyon wrens with us for some time. The bird's song sails effortlessly down from on high and continues out across the desert where it eventually disappears without a trace. It leaves no physical mark on its world, it needs no excuse, no justification at all.

Death in a saguaro forest

A grazing cow leaves a battery of physical marks on its world, directly with its hoofprints, cowpies and vast appetite, and indirectly via its gun-toting, lion-killing caretakers. While converting forage into attractive termite food, cows have removed a great deal of Western grass cover, probably facilitating the spread of relatively cow-proof prickly pear cacti, a change very much to the liking of our "native" pigs. But although termites and peccaries may be advantaged by grazing cattle, desert tortoises evidently are not because they prefer to eat some of the very same plants that cows are so good at harvesting. Cattle grazing may have, therefore, brought about diverse changes affecting an elaborate network of organisms, one that probably includes our most famous cactus, the saguaro.

A large, healthy saguaro with its arms raised in an almost human salute looks so utterly confident of its importance, so big and tough, that it has the appearance of a permanent fixture in its landscape. And yet one day the saguaro lies collapsed on the ground, its arms thrown out in front of it, like a person suddenly dead from a massive heart attack. Or if not wind-thrown into instant death, the saguaro may develop a small but ugly brown splotch of decay, which grows faster and faster until in a matter of a few weeks, the plant's huge body has literally disintegrated under its green skin. At this stage, the exterior of the cactus will be marked with rivulets of rotting exudate, perhaps an arm or two will have fallen off, while its internal flesh has undergone an insidious conversion into black soup. Later the dead cactus stands skeletonized, with its now brown skin fallen down at the base of its trunk like a baggy sock drooped around an ankle. What was once a monument to symmetry, larger than life, is now uneven, awkward, and painful to observe, a slap-in-the-face statement about the mortality of living things.

It has seemed to me that I meet dead and dying saguaros far too frequently these days, and I am not the only person with this worry. For years now, biologists and nonbiologists alike have known that the cacti in the Saguaro National Monument near Tucson are dying at a distressing rate, so fast that it appears unlikely that the forest will persist much longer in its present form.

This federal preserve was created in 1933 to honor and protect a remarkable stand of thousands upon thousands of giant saguaros that occupied the foothills and bajadas near the Santa Catalina Mountains. In their book, *The Changing Mile*, James Hastings and Raymond Turner include two photographs of a portion of the forest, one shot in 1935 and the other in 1960. The two photographs were taken from exactly the same location, and by comparing them one can see that one-third of the adult saguaros had collapsed in just twenty-five years. Researchers working in the forest in the sixties projected its complete disappearance by 1998, given the apparent mortality rate coupled with the evident failure of young seedling saguaros to take the place of those that were dying.

But why were the adult plants dying in such numbers during the period from 1935 to 1960? Desert biologists offered several possible reasons. Stanley Alcorn and his co-workers believed that the adults were being infected by a virulent bacterial disease that spreads from victim to victim, causing an epidemic of deaths from internal decay. Charles Lowe and his co-workers disagreed, arguing that the bacteria merely scavenged cacti that were dying or had already succumbed to other causes, such as severe freezes. Hard freezes of twenty-four hours or more occur occasionally in the Tucson area, and one particularly strong cold snap in mid-January 1962 killed great numbers of adult and young saguaros alike.

Lowe attempted to test the bacterial necrosis hypothesis by wounding hand-reared cacti and placing rotting material from infected specimens in the wounds. He reported that the healthy saguaros resisted infection, as he expected, given his conviction that freezing weather was the real killer.

Stanley Alcorn, however, performed similar experiments of his own with diametrically opposed results. He succeeded in transmitting lethal doses of the bacteria in question from dying to undiseased saguaro arms, kept alive in a greenhouse for years until infected. His work revives the possibility that the bacteria is an active killer of the cacti, not merely an exploiter of dead or dying specimens that have been victimized by other agents of saguaro mortality.

The death of the attractive mature adults, by whatever means, was bad enough, but what equally disturbed saguaro enthusiasts was the absence of young replacements, which appeared to spell long-term disaster for the saguaro population at the monument. Here, too, there was no shortage of possible explanations for the absence of youngsters, among them the grazing hypothesis, which is where cows enter the picture.

Although it would seem that the federal government would accord a National Monument full protection from the livestock industry, ranchers complained so bitterly when the monument was set aside that the feds permitted them to run cows throughout the entire area until 1958—despite reports that the range was in terrible condition due to overgrazing. Federal authorities even increased the number of cattle permitted on the Tanque Verde allotment from 269 to 555 shortly after the formation of the monument. And when they reversed the policy of grazing in 1958, they did so only partially. A large chunk of the monument was grazed legally until 1978. The monument managers did not manage to track down and shoot the last feral cow trespassing on the land until one fine day in 1985.

According to some observers, cows are not compatible with the establishment of saguaro seedlings, which they destroy in a host of ways, although direct consumption of young saguaros is the least of their effects. More damaging is the understandable tendency of cattle to seek the shade of a mesquite or paloverde during the heat of the day. Most young saguaros get a roothold in the shelter of desert trees and shrubs, which provide shade in the summer and protection against freezes in the winter. A baby saguaro is no match for a five-hundred-pound cow, should one of these beasts choose to step on it or lie down upon it or even flatten it with a deposit of heavy, wet dung.

Perhaps the worst thing that cows do for saguaro seedlings is something that they do frequently and well, which is to consume desert grasses. These plants provide low-lying cover that complements the microclimatic protection offered by "nurse" tree canopies. Grasses conceal the very small cacti, which do have many native consumers, including jack rabbits and small desert mice and rats. In addition, grasses blanket young saguaros, protecting them against extremes of heat or cold, ameliorating the effects of damaging temperatures.

The discovery that there were no baby saguaros to speak of in the monument was made in the early 1960s, right after the removal of cattle from the premises. By the 1970s, in the areas where cows were on their way out, grasses were on the rebound and saguaro seedlings had begun a comeback. In his study of immediately adjacent sections of the monument, one that had been grazed until 1958 and the other until 1978, Fareed Abou-Haidar found that the age structures of populations of saguaros in the two areas differed in an instructive way. In the region where grazing had ended twenty years earlier, there were more than six times as many young saguaros in the eleven-to-twenty-year bracket (as estimated by their size) compared to the plot next door where grazing had continued until 1978. Abou-Haidar's results support the hypothesis that cattle grazing indirectly destroys young seedlings, creating a population that becomes older and older, and eventually fewer and fewer as death claims the mature saguaros one by one. Amazingly, Abou-Haidar's study appears to be the only attempt actually to test the hypothesis that cattle grazing has negative effects on saguaro seedling recruitment. The ratio of speculation to scientific test is disturbingly high when it comes to understanding modern population dynamics of saguaro cacti.

Nevertheless, the fact that an end to grazing apparently permits young saguaros to join the population once again offers hope for the future of the monument. But in the meantime, the large adult saguaros there continue to die off with the sad result that only a handful of mature, photogenic saguaros with many arms will persist in the eastern portion of the monument by the turn of the century. Raymond Turner and others believe that the total population there will be as great or greater than it was in the 1930s, but that the age distribution of the cacti will be utterly different. Instead of a preponderance of ponderous adults, the vast majority of saguaros will be immatures without arms, too young to produce flowers, too small (only head high to a human or smaller) to inspire the kind of awe and admiration that we rightfully accord an old desert icon with fifteen massive arms and a track record of survival for 150 years or more.

The changes that are occurring in the Saguaro National Monument can, therefore, be laid in part at the feet and mouths of cattle (with even greater blame for the humans that permitted ranchers to dictate management policy at the monument). Grazing pressure probably contributed to the elimination of entire cohorts of potential recruits into the local population. The impact on the saguaro community was not obvious at first, but over decades the result may be a monument to the capacity of man and cow to destroy even those desert places that we intended to preserve.

Randolph Canyon and Burro Creek

It is gratifying to hear that the saguaro population in Saguaro National Monument is already beginning to recover despite decades of grazing abuse. The same resilience of desert communities is apparent in Randolph Canyon, which lies within the roughly 10 percent of Arizona's public lands that are now officially wilderness. I have been out to the place many times, but for some reason, most of my hikes have been made during times of drought when the landscape included dry streambeds, skeletonized mesquites and churned sandy soil littered with fossil cow dung—and the occasional dead cow. Cows may seem to be an unlikely feature of a true wilderness experience, but the congressional architects of the wilderness designation succumbed to the potent livestock lobby, just as had the designers of Saguaro National Monument. The solons decreed that ranchers could still run their cattle in official wilderness areas provided that they could get them in and out without recourse to roads.

On today's hike into the Superstition Wilderness, I am surprised and delighted to see no cows at all in Randolph Canyon. Are we in the twelve-month cattle-free interval for the Millsite Grazing Lease? Or has the abundant water and greenery enabled Mr. Martin's cows to spread out across the land and stay out of view during my stroll through "their" terrain? Whatever the reason, I am grateful for their absence.

Thanks to a recent rain, which was only the last in a series of late winter and early spring storms, water now flows along great stretches of the stream in the canyon, not puddling up in a few polluted tanks. The desert plants today bear no resemblance to the drought-deprived specimens of the past. Dozens of hedgehog cacti are in bloom; their raspberry red flowers compete for attention with the canary yellow blossoms of the many prickly pears. Cacti are said to be closely related to the rose family, and this spring's extravagant floral display makes the argument completely persuasive.

Poking up among the sturdy, spine-covered hedgehogs and prickly pears, the Mariposa lilies offer a stunning contrast in design to the many-petaled cacti flowers. Each lily is topped with one or two extravagantly elegant, bright yellow or deep orange-red cupped flowers. The tuliplike flower balances on the top of thin stems more than a foot long. The ephemeral delicacy of the lilies is enhanced by the plant's nearly complete absence of leaves. The lilies, top-heavy with their flowers, lean to one side or the other.

Two have twisted up through the leathery green leaves of an agave. Their red flowers lie cradled between the spiny borders of neighboring agave leaves.

Along a sandy creek, Bell's vireos crank out their odd chortled song over and over. Migrating black-throated gray and Townsend's warblers search through desert trees for insect food that will sustain them on their travels. A beam of sunlight slips through an opening in the thin foliage of a mesquite. A Townsend's warbler flits into the shaft of sunshine; the warm yellows on its face come alive with light before the bird dives back into the liquid shade of the tree.

Over the Red Tank pools, big orange dragonflies cruise repetitively back and forth. Two males pull into formation, like fighter planes, one slightly ahead of and above the other. They fly off in a straight path, gaining altitude steadily as they head downstream and around a bend of twisted rocks that descend to the water.

A canyon wren comes skipping down from a polished rock wall to forage in the tangled debris along the edge of the stream. It peeps in alarm when I push between some saplings near the water, but the wren does not flee as I stop in my tracks. Instead it resumes its search for flies and other small invertebrates along the edges of rocks and water-soaked twigs and branches. The wren ducks as a dragonfly darts over its head. Slowly working the edge of the stream, it plucks tidbits and bobs ever closer to where I stand with one foot on a rock in the water and the other on the shore. Soon it passes directly between my legs and continues downstream, concentrating solely on what is right in front of its curved black bill.

In this water-rejuvenated desert, the recovery of living things announces itself everywhere. Even in places where they cannot be seen, creatures exploit the temporary abundance of water. The little stream in the canyon flows erratically in places, diving beneath the sand for forty feet before resurfacing to slide along above ground for awhile, only to return to a wholly subterranean aqueduct again. In among the buried sands, the flowing water sustains a weird menagerie of small invertebrates, among them creamy white amphipods more blind than bats. The amphipods have no need for eyes because they live their entire lives deep within the fine gravel far underground in constant blackness where hidden water inches slowly downstream.

The microfauna of subterranean stream flows have to cope with dramatic changes in their boom or bust environment. The amount of sand and gravel serviced by water that is available for colonization by these creatures fluctuates annually, as the amount of water coming from the watershed increases and declines. After a severe summer storm, the sudden surge of a flash flood can scour great stretches of desert streams, flushing whole ecosystems downstream, turning the gravel beds upside down and depositing them elsewhere. The desert stream invertebrates in the sand column somehow find refugia amid the chaos and recolonize the available, reordered habitat after the flood subsides.

The same floods that challenge the wet sand animals in their hidden world can do a number on the living things that inhabit the visible world of the streamside. Up on Burro Creek in western Arizona, unusually heavy winter rains produced a massive flood in 1979 that ripped the vegetation from the broad plain through which the normally placid creek traveled. Boulders and sand replaced mesquites and grasses.

Jerry Brimhall and his father, Lee Brimhall, have grazing rights along the Burro Creek floodplain, which is administered by the Bureau of Land Management. The Brimhalls continued to run their cattle on the devastated streamside after the flood. By 1981 the BLM was concerned enough about the way things looked to perform an environmental-impact study on the effects of cattle on the recovery of the riparian zone. They decided that the Brimhalls would have to reduce the number of cattle that foraged on BLM land.

The Brimhalls were not amused. They enlisted the aid of Senator Dennis DeConcini (D–Arizona), who intervened on their behalf with the BLM. It is standard practice for western Senators to perform this service for their rancher constituents, one of the more notorious examples involving Senator Stephen Symms (R–Idaho) and a Forest Service Ranger named Donald Oman. Mr. Oman's assignment was to keep tabs on the cows that graze in the Sawtooth National Forest in Idaho, where he took it upon himself to count the cattle there to insure that ranchers did not exceed their permit levels. And he expressed some strong feelings about the range he patrols, which were encapsulated in his claim, "This land belongs to 250 million Americans, not just the guy with the cattle-grazing permit."

Those were fighting words as far as the local cattlemen were concerned. One of the ranchers, Winslow Whitely, was willing to go on the record suggesting that Mr. Oman get out of the Sawtooth "or he's going to have an accident." This statement sounded very much like a threat to inflict bodily harm on Mr. Oman to the New York Times reporter Timothy Egan, who chatted with Mr. Whitely in the summer of 1990 (not 1890).

"Yes, it's intentional," confirmed Mr. Whitely. "If they don't move him out of this district, we will."

However, before putting out a contract on Mr. Oman, the public lands graziers first dropped in for a chat with Senator Stephen Symms. Shortly thereafter, Mr. Oman received word from his Forest Service superiors that he was on the way out of his Sawtooth district.

Mr. Oman contested the decision from on high, using the federal whistleblower's hotline to register his complaint at being forced from his job for purely political reasons. Federal investigators issued their whistleblower's report on the Oman affair in the fall of 1990. They had some criticisms both for Mr. Oman and his superiors, but in the last analysis, Ranger Oman was told by his bosses that he would not be forced to accept the transfer that the Forest Service had earlier arranged for him.

No doubt Mr. Whitely is not pleased with the way in which he and his fellow ranchers have been thwarted—thus far. He can perhaps content himself with the knowledge that, even with the zealous Mr. Oman operating on "his" turf, he is in no danger of immediate poverty. Mr. Whitely's 1990 grazing permit gave 1,563 of his steers access to public lands. By his own reckoning, each animal yielded about a $250 profit, generating a total of roughly $400,000 in private gains from public lands. In stark contrast, well over one-third of the public range of the Sawtooth was placed in the "poor condition" category by the General Accounting Office when they surveyed the region in 1988.

In any event, when the Brimhalls went to Dennis DeConcini, they were following a time-honored tradition, and in their case the result was more traditional than it was for Winslow Whitely. Agency officials at the BLM decided that they would drop the requirement that the Brimhalls run fewer head of cattle on Burro Creek. But they did impose a pasture-rotation plan on the ranchers so that the stream areas had intermittent relief from constant grazing and pounding. In 1990, a reporter for the *Arizona Republic* enthused, "The land that was covered by nothing but rocks and boulders 10 years ago once again is green. Deer, javelina, eagles, beavers, hawks and an occasional mountain lion roam amid the growth. And somewhere, leaving behind little evidence of their presence, are 150 head of cattle."

I am skeptical of the claim of near invisibility for 150 head of cattle along Burro Creek, but it is encouraging that desert vegetation can stage a substantial comeback in just a decade in a place with that many steers at work. In fact, the recovery seems close to miraculous, although my experience in Randolph Canyon suggests that the desert can repair itself remarkably quickly, if given half a chance. And Lee Brimhall had the good grace to acknowledge that "Mother Nature had something to do with it, too" in addition to the BLM's DeConcini-induced management plan.

Mother Nature, in the form of desert-adapted plants and animals, probably succeeded on Burro Creek *despite* the BLM and the Brimhalls. A study of a Gila River mesquite bosque by Wendell Minckley and Thomas Clark suggests that the riparian forests of southern Arizona may be constantly torn down by floods and then recreated as surviving trees send out propagules to recolonize new terraces formed by the river. As the new trees grow, their presence affects the deposition of more material from the nonflooding river, creating more habitat that mesquite can invade. The new cycle of forest expansion ends when another severe flood sweeps down the river, cutting into the riverbank, sweeping mesquites into the water, piling boulders onto the floodplain. On the upper Gila River there have been at least twenty potentially habitat-altering floods of this sort in the past century.

Perhaps severe floods have occurred less frequently on upper Burro Creek, which drains a drier region than the Gila River. But if I were Lee and Jerry Brimhall and the BLM, I would not be completely surprised to see another ferocious flood come surging down the creek to undo the past handiwork of Mother Nature. When that flood happens, it will create another opportunity for Her, not the BLM, to go to work again. If I were Mother Nature, I'd be grateful to have a chance to do my job in the absence of Lee and Jerry Brimhall's hard-eating, grade-A, Arizona steers.

The masked bobwhite rides again

Although many elements of the desert have the capacity to bounce back on their own after adversity, there is a limit to the resilience of the inhabitants of this world. We have, for example, pushed the masked bobwhite right to the edge.

Back East, bobwhite quail are in no danger of extinction. They are an integral part of many southeastern environments, free to make an impression on those lucky enough to live around them. The riot of noise when a covey bursts in all directions from underfoot. The scuttling of a single bird scampering down a row in a cornfield. The signature of summer, a distant *bob-white*, drifting up from a far-away hay pasture on a hazy, hot afternoon.

Herbert Brown, a native of West Virginia, did not expect to experience these sensations while in Sonora, Mexico, at the turn of the century—but he did. "It is not easy to describe the feelings of myself and American companions when we first heard the call *bob white*. It was startling and unexpected, and that night nearly every man in camp had some reminiscence to tell of Bob-white and his boyhood days. Just that simple call made many a hardy man heart-sick and homesick."

Brown had come across masked bobwhites in Arizona in previous years, but by 1904 he knew that he would not again hear the call of the bobwhite in that state. He noted, "For the past several years [the bobwhite] has been safeguarded by law in this Territory [of Arizona], but unfortunately there are none left to protect." The last two masked bobwhite seen in Arizona became museum specimens on December 29, 1887. Their skins eventually made their way to museum drawers in Flagstaff and Tucson, where for many years they provided the only remaining trace of the little quail in the United States.

Brown was something of an ornithologist, and he made the first scientific description of the masked bobwhite, which he assigned to the same species as the familiar gamebird back East. The masked bobwhite had lived in isolation from its other relatives long enough to achieve certain differences in coloration, notably a dark chestnut breast and black throat and head for the male in place of the white throat of other bobwhite populations. But the separation had neither been long enough nor complete enough to erase many similarities between the masked bobwhite and bobwhites living elsewhere. The females of the various populations are essentially identical. Moreover, the call of the masked bobwhite (a loud whistled bob-white!) never diverged from that of other bobwhites. Thus, persons familiar with the bobwhite of the eastern United States, such as Herbert Brown, could immediately identify the bird when they heard it calling in southern Arizona or northern Mexico.

Although other ornithologists initially felt that the bird deserved to be placed in a species of its own, eventually (in the 1940s), the ornithological community concluded that Brown was right. They agreed that the masked bobwhite should be placed within one large diverse species, rather than split off as a distinct species in its own right. The disappearance of the masked bobwhite from Arizona therefore represented the loss not of a full species but of a distinctive subspecies or race of the very widely distributed bob-white.

Still, even though it was just a subspecies and even though its range embraced parts of northern Mexico (where less than 500 individuals currently persist), the masked bobwhite was an attractive addition to the avifauna of the United States. The utter extinction of the masked bobwhite in Arizona is all the more sad because once upon a time it was probably a fairly common bird here, although restricted to a small part of the state. Prior to the turn of the century, it occupied plains and river valleys in southcentral Arizona in places with a dense cover of native grama grasses. Then it was gone.

The cause of its remarkably rapid disappearance has been traced with reasonable certainty to the stupendous overgrazing that took place in the 1880s and 1890s. The masked bobwhite's preference for grassy plains, a habitat limited to extreme southcentral Arizona, brought it into direct conflict with the Texas steers driven into this country. The steers won. The million plus cows in Arizona before the turn of the century literally ate the quail out of house and home, just as current overgrazing in Sonora has all but finished off the Mexican representatives of this race. Again, here is Herbert Brown's account of the effect of drought and overgrazing in Arizona.

"During the years 1892 and 1893 Arizona suffered an almost continuous drouth, and cattle died by the tens of thousands. . . . The hot sun, dry winds and famishing brutes were fatal as fire to nearly all forms of vegetable life. Even the cactus, although girdled by its millions of spines, was broken down and eaten by cattle in their mad frenzy for food. . . . I saw, later, what I had never expected to see in Arizona, Mexicans gathering bones on the ranges and shipping them to California for fertilizing purposes. I have thus particularized, for in these dry bones can be read the passing of the Partridge from many a broad mile of the Territory."

Included among the "Partridge" that Brown referred to were several species of quail, not just the bobwhite. All were hit hard by overgrazing and overhunting. According to Brown, a law designed to protect all species of quail was introduced into the territorial legislature in 1887. But because Gambel's quail were regarded as pests "by the ranchmen in the Salt River valley [the Phoenix area] . . . the legislators from Maricopa County threatened to kill the bill unless the clause protecting "Quail" was stricken out." Then, as now, ranchers displayed an interest in the legislative process.

Despite the delay in instituting legislative protection for the several species of quail in Arizona, only the masked bobwhite succumbed completely, presumably because of its limited distribution in the state and absolute dependence on dense grassy habitats. Other birds that were heavily dependent on grasslands also suffered long-term declines, although not total extinction. Baird's sparrow, a magnificently nondescript little sparrow, lives only in grassy plains in both its winter and summer grounds. The several ornithologists active in Arizona prior to the 1880s found it a common wintering bird in the appropriate habitat in southeastern Arizona. But only three specimens were collected in the decade beginning in 1880, despite the fact that Arizona was apparently crawling with naturalists eager to explore the area and collect its unique avifauna.

Incidentally, one of these early naturalists was an army man, Major Charles E. Bendire, now immortalized by Bendire's thrasher. Herbert Brown reports, "Prior to 1870, but just when I cannot now say, Major Bendire, then a Lieutenant of Cavalry, was stationed at Camp Buchannon, on the Sonoite, almost in the very heart of the country where the Bob-whites used to be, but, oddly enough, he did not see or hear them. At that time the valley was heavily grassed and the Apache Indians notoriously bad, a combination that prevented the most sanguine naturalist from getting too close to the ground without taking big chances of permanently slipping under it. For many years Indians, grass, and birds have been gone."

Unlike the masked bobwhite, Baird's sparrow has persisted in Arizona, perhaps because it is only a winter visitor and migrant here, going on to the plains of the northern United States and southern Canada to breed. The sparrow seems able to tolerate some grazing in the prairies it calls home. The masked bobwhite could not, and it had nowhere to go.

But the bobwhite did have some persons dedicated to its well-being. Since 1937 various attempts have been made to reintroduce the bird into Arizona. Initially, J. Stokely Ligon captured wild birds from Mexico and released them in assorted spots in Arizona and New Mexico that he judged contained passable habitat for the birds. All Mr. Ligon's efforts failed, perhaps because the sites he selected for the transplants were too different from their home habitats. All his release points lay far outside the presumed historic range of the masked bobwhite.

Subsequent efforts to bring the birds back have focused on using captured wild birds as breeding stock to create populations of captive-reared birds for eventual release into the field. The early attempts to produce new stock were severely hindered when vandals broke into breeding pens and destroyed many captive birds. *The Birds of Arizona* summarized the outcome of all these heroic efforts as of 1964. "Attempts at reintroduction have been unsuccessful, as there is no ungrazed grassland within the former range [of the masked bobwhite] within Arizona."

Subsequently, the U.S. Fish and Wildlife Service took over the task. They combined a small number of pen-reared birds with another infusion of wild masked bobwhites captured in Sonora, Mexico, to form a new breeding stock. Thousands of descendants of these birds have been released over the years onto grasslands in southern Arizona. But nobody said it would be easy to get the masked bobwhite back in the saddle again in Arizona. At first, only the local coyotes had reason to be enthusiastic about the "recovery" program. Pen-reared birds are simply too naive to hack it on their own in the real world.

After the failure of the early release program, the USFWS developed a new technique in which the captive-reared birds were provided with wild-caught foster parents (sterilized male Texas bobwhites seem to do the best job in educating their adopted offspring to the dangers of life on the range). This strategy and others resulted in some success so that by the 1970s a population of about thirty trained birds had survived their first winter in the wild and had produced a few young of their own the subsequent summer.

Continued releases built the population of calling (reproducing) males up to seventy-four by 1979, but then disaster struck, once again in the form of drought and grazing. The birds' home was the Buenos Aires Ranch near Sasabe, Arizona, a hamlet on the Arizona-Mexico border. The land that the Fish and Wildlife Service leased from the owners of the ranch was grazed continuously during the 1979 drought, with the result that cows gobbled up the grass cover that the birds find absolutely essential. By 1983 only a few males remained to call in the summer breeding season.

Recognizing at last that cattle are incompatible with masked bobwhite, the USFWS purchased the Buenos Aires Ranch in 1985 for $8,900,000, after which it took the dramatic and controversial step of getting cows completely off the land before stocking it with captive-reared bobwhite. The Arizonan ranching community protested loudly.

Happily, these protests did not sway the USFWS, which acquired slightly more than 21,000 acres of prime grassland outright while leasing an additional 90,000 from the Arizona State Land Department. Ultimately, these state-owned lands are to be transferred to the federal government in exchange for parcels of federal lands elsewhere in Arizona. When this happens, Wayne Schifflett, the manager of the Buenos Aires National Wildlife Refuge, will be in complete charge of by far the largest block of cattle-free prairie in the state. The total investment of $10 million in the masked bobwhite program to date is justified by the revitalization and protection of a superb grassland ecosystem, no matter what happens to the bobwhite over the long haul.

In the summer of 1989 there were 50 calling males, but when the monsoon rains failed, it seemed probable that the Buenos Aires bobwhites had had it. Steve Dobrott, the Fish and Wildlife biologist at the refuge, held out little hope for the few birds that had hung on over a series of drought years. On top of everything else, the winter of 1989–1990 was also dry, so that Steve was prepared for the worst when the 1990 breeding season rolled round. But somehow 40 calling males put in an acoustical appearance in the summer of 1990, and then the monsoon season came. This time, it wasn't a bust. With 14 inches of welcome summer rainfall, the perennial grasses ran riot for the first time in years, providing abundant food and cover for the survivors of past releases and the approximately 600 new birds that were given their freedom. Steve now has high hopes for the quail, at least over the short run.

Drought years will be back again sometime and the masked bobwhite is not out of the woods and onto the prairie yet. A single small population is hardly an ideal prescription for long-term survival of the bird in the United States. Conservation work on the masked bobwhite's behalf proceeds in Mexico, but that country possesses no federally protected reserve for the bird and overgrazing has destroyed virtually all of the bird's habitat there. One hopeful sign is the enthusiasm of one young Mexican rancher for the masked bobwhite. This person has recently assumed responsibility for managing his family's estate, and he is deeply interested in the preservation of a bird that is part of his childhood experience. Perhaps his land holdings will serve as one safe haven for the masked bobwhite in Mexico.

Meanwhile, down on the refuge the native grama grasses have formed a luxurious, wind-rippled carpet anchored in many places by dark green mesquite trees. Prickly pear cactus laden with scarlet fruit and barrel cacti ornamented with orange-red flowers poke up through the swaying grasses. Horned larks scamper across the dirt tracks and meadowlarks coast over the swales. Pronghorn antelopes, another species that had been extirpated from the area but are now successfully reintroduced, wander through the knee-high savannah. Cassin's sparrows leap into the air, uttering their buzzy trilling song as they descend into the concealing, sustaining, beautiful grasses.

On the western horizon, the Baboquivari Mountains run north and south, abruptly delineating the grassland ecosystem. As evening comes, a distant male masked bobwhite whistles one last explosive *bob-white!* into the wind that combs the uneaten grama grasses. It is a hopeful song in an Arizona prairie that has been permitted to demonstrate its power of recovery, its capacity to sustain a rich diversity of life, its ability to preserve a history worth preserving.

DESERT HOPE

In calling up images of the past, I find that the plains of Patagonia frequently cross before my eyes; yet these plains are pronounced by all wretched and useless. They can be described only by negative characteristics; without habitations, without water, . . . they support merely a few dwarf plants. Why then, and the case is not peculiar to myself, have these arid wastes taken so firm a hold on my memory? Why have not the still more level, the greener and more fertile Pampas, which are serviceable to mankind, produced an equal impression? CHARLES DARWIN, *The Voyage of the Beagle*

Life in a saguaro forest

A quiet, overcast morning with no trace of a breeze. A residuum of humidity from last night's monsoon shower, which laid the dust and perfumed the hothouse desert air. Silence barely broken by the sound of disturbed gravel slipping down a steep incline on Usery Peak.

A Gila monster edges across the slope in its flat-footed, legs-far-apart waddle, sending gravel trickling downhill. The top of its head is covered with a brown patch of dirt probably acquired during an earlier inspection of a rodent burrow or rabbit den. The lizard's tail is thin, testimony to the hard times that have forced it out to look for food in July when it should be tucked safely away in its own underground bunker, waiting out the dog days of summer. The monster reaches a maze of big boulders and lurches in among them, removing itself from view.

I carry on to the upper ridge on Usery Peak. I could choose to look on the western horizon where the skyscrapers of downtown Phoenix are barely visible amid a miasma of car exhaust, or I could keep my eyes on the local terrain. I choose to focus on the local terrain, admiring the magnificent stand of big saguaros that occupies the south-facing slope of the peak. The many visible cacti here mostly range from armless totem poles about my height to elegant saguaros over twenty feet tall and arrayed with various curved arms, each cactus with its own distinctive shape and personality. The very largest individual in this population gives the illusion of being something of a skyscraper itself, its long graceful arms arcing up in a hugely triumphant way. Earlier in the year, a brown towhee had a nest in a big jojoba neighboring the largest saguaro; the cactus towers over the shrub, pushing it into visual insignificance.

The current champion of size and symmetry among the saguaros, however, has only recently become the largest and most impressive specimen among the dozens that stand at attention up and down the mountainside. It assumed its dominant position upon the death of an even larger cactus just a few years ago. Nor was the demise of this other individual unique in the years since I have been coming to Usery Peak. In fact, an uncomfortably large number of the magnificent giants in the area have gone down over the last ten years, creating the impression of a population in severe decline.

Although it is likely that the troubled future facing the saguaros occupying Saguaro National Monument has been largely dictated by the actions of humans and domestic livestock, changes in the Usery Mountain saguaro population cannot be assigned to the "unnatural" cause of cactus-tromping cows. The saguaro population on Usery Peak has never had to endure the company of a herd of steers. The hillside is far too steep and rocky to have been negotiated by even the most adventurous of cows. The ascent and descent pose a substantial challenge for humans, let alone a rancher's meal ticket. I did once find the skeleton of a horse about halfway up the peak, its leg bones scattered among the boulders, its fleshless skull a redolent trophy for transport to my suburban backyard. How the animal managed to reach its ultimate resting place is a mystery, and I cannot imagine that many other domestic animals have attempted the feat.

Nevertheless, even without cows, the peak has a markedly different look to it now, compared to just ten years ago, something that I verified by examining photographs of the site taken ten years apart. My 1980 slide contains forty-seven visible cacti, ten of which are magnificently large specimens with the array of upraised arms that gives big saguaros their immense aesthetic appeal. A photograph taken in 1990 reveals that seven of the forty-seven saguaros are missing. Among the deceased are five of the armed cacti of 1980, including the two largest individuals on the peak. Even the two unarmed specimens that bit the dust were among the tallest members of the population in 1980.

The skeletons of those that have died are still very much in evidence on the hillside. The durable internal ribs lie exposed, bent and twisted, fleshless and brown, all that remains of the once imposing cacti that toppled onto the boulders beneath them.

Because I was aware of the deaths of the larger, older giants, I had formed an impression of a depleted hillside, a forest with major lacunae. But when I counted up the cacti visible in the 1990 slide I found nine armed saguaros, just one less than in 1980. Over the past ten years, several of the totem-pole cacti in 1980 have acquired arms of their own. Admittedly, the newcomers to the armed saguaro population are not nearly as majestic as the multi-armed monarchs that went down, but at least they are there to grow, and someday they may truly replace the major cacti that have disappeared in recent years.

Moreover, in the 1990 photograph, I can see six small cacti (which nonetheless are surely more than ten years old) that I cannot find in the 1980 slide. Obviously, these specimens were smaller still in 1980, a year when the hillside vegetation was relatively lush and could easily have concealed baby saguaros from my camera. Still smaller saguaros exist that are not visible in either photograph. Therefore, it is entirely possible that today's Usery Peak population of these marvelous cacti is the same as or greater than it was in 1980, despite the loss of its most notable and conspicuous members during the following decade.

Nor is Usery Peak unique in the apparent numerical health of its population. In their studies in the Pinacate region of northwestern Mexico and elsewhere in the Sonoran Desert, Hastings and Turner have documented that saguaro cacti populations can change substantially over the years, even in areas that have not been heavily grazed. In some places, saguaro numbers are on the rise, increasing every bit as dramatically as the mature cacti in the Saguaro National Monument have been plummeting toward oblivion in recent decades.

I suspect that those saguaro populations that are falling because of *natural* causes may be ones in which a large proportion of the adult population belongs to a single cohort of oldsters, a cohort that started off during a rare short-lived period long ago when conditions happened to be especially favorable for the survival of saguaro seedlings. In such populations, there will come a time when the adult saguaros of this year-class are still numerous, robust and multi-armed, creating a relatively dense forest. But as they grow older still and begin to succumb to the diseases of old age, the bad luck of a hard freeze, or an unusually fierce wind storm, their deaths will create multiple gaps in the forest. Their relatively rapid disappearance alters the scene over a short enough span so that human beings can see for themselves the disturbing impermanence of all living things.

For a while longer, however, there are sufficient saguaros to take the place of those that tumble down Usery Peak. Here the cacti have the look of survivors. Perhaps the population will persist for a very great time, even after humans and their camp-follower cows have disappeared from the planet, with one generation of saguaros gracefully giving way to the next until a new Ice Age pushes them slowly off the mountain and replaces them with junipers, agaves and beargrass.

The black bear in Ballantine Canyon

The Forest Service trail through Ballantine Canyon climbs up and up, and up some more. Weaving its way, zigzagging higher, it leaves the Sonoran Desert behind. The last saguaros stand amid a great jumble of gray boulders on a south-facing slope at about the five-mile mark. Down the trail from this vantage point, the rocky terrain sports a spartan cover of mesquites and acacias, enlivened with golden patches of teddy-bear chollas and the occasional flourish of an upright saguaro. From this point on, a chaparral composed of oaks and junipers and manzanita blankets all but the most prominent of the rocky outcrops that burst through the leathery green leaves of the stunted forest. Here and there a clump of beargrass or a starburst of blue-gray agave leaves accents a landscape that no longer qualifies as desert.

At the six-mile mark, the foundations of a handful of prehistoric rock-walled buildings constructed sometime prior to the 1450s lie across the trail. A red potsherd peeks out from under a pad that has fallen from a partly eaten prickly pear cactus. The cactus battles the local acacias for ownership of one corner of a long-extinct household.

The stream that once served the handful of Indians who lived here half a millennium ago flows in a tentative fashion. Lower down, closer to the trailhead, the wash is empty, choked on sand and gravel deposits. Grooves etched in the rock sheet that forms the streambed there speak of ancient winter rains, when water rushed along, scraping out souvenirs of its passing.

The trail, which had been following the gently ascending creek for some time and therefore had been gaining altitude at a reasonable rate, now moves off to run up an imposing canyonside at an imposing angle. A gang of acacias lies in wait to scratch and prick the trail user, but eventually the thorny acacias give way to a dense stand of manzanita, a gorgeous chaparral plant with deep cinnamon red trunks and limbs. In its season, now past by about a month, manzanita also produces a bonanza of little reddish fruits. There on the trail lies first one and then another and then many, many more bear scats filled with bits of manzanita fruits and seeds. The scats are chunky, big and hearty, but they are now also dried and fragile, scattering into a hail of fragments when kicked. The black bears in Ballantine Canyon obviously had a field day during the manzanita harvest. I am pleased that the harvest is long over because my one-man primitive campsite is to be up the trail only a short distance now, at Rock Tank, where ponderosa pine trees replace the chaparral. I do not relish the prospect of sharing my camp with a bear.

Rock Tank contains considerable water within its sculpted hard rock tinajas. The stream now flows with appreciable force, slipping from one pool to another before tumbling down a rocky incline into Ballantine Canyon. I walk around the area looking for bear sign, and finding none, I locate a level place on which to place my expensive backpacker's self-inflating mattress and my Recreational Equipment Inc. sleeping bag. The sun begins to descend behind the mountains far to the west. A yellow glow suffuses the sky with a melancholy aura. I eat a granola bar and absorb the mood of the evening, then walk around a large rock on the way back to my sleeping bag. On the other side of the rock, twilight offers sufficient light to make out a handsome prickly pear cactus with a few ripe fruit projecting from its pads.

By the cactus lies a great mound of moist bear dung, dark red from the purple fruits the bear ate not so long ago. The first star of the evening finds its place in the night sky. My heart beats a little faster. An erratic gust of wind scuttles through my U.S. Forest Service–subsidized campsite, which has now acquired an extra element of wildness that it did not have before.

The Mazatzal Wilderness Area

Not far from Ballantine Canyon, the long, red eastern wall of the Mazatzals begins its run north, a wall that carries on for miles, broken only by the occasional deep canyon cut into the mountains. Behind the wall the mountains and valleys follow one on the other for as far as the eye can see, comprising a great sweep of central Arizona that since 1964 can call itself the Mazatzal Wilderness Area. It is true that all the springs within the 200,000 acre wilderness were long ago found and prosaically named: Lower Sheep Spring, Sheep Creek Seep, Horse Camp Seep, Brody Spring, Jones Spring. It is true that there are ten grazing leases that impinge on the wilderness and one, the Bull Springs Allotment, that lies totally within the Mazatzal Wilderness Area. It is true that the Forest Service advises, "If you wish to avoid livestock, check with the local ranger for areas that will be free of stock during your visit." It is true that mineral exploration is permitted in the Mazatzal Wilderness Area.

But at least no sheep, steer, horse, rancher, miner or backpacker has been on this trail since it snowed four days ago. The half-foot to a foot of crystalline snow that receives my footprints contains an ample record of previous pedestrians, not one of which was a fellow human or domesticated beast. A long line of deer tracks alternates and coincides and alternates with the footprints of a coyote. A flurry of rabbit prints crosses the trail at right angles. A squirrel with muddy feet has left its faint marks barely outlined in a short run over the crusty surface. A rufous-sided towhee scoots from one patch of manzanita chaparral to another, its white belly and white outer tail feathers harmonizing with the snow cover, its rufous sides color-coordinated with the red-barked manzanita, its black back camouflage for the shadows pooled beneath the tangled scrub.

The trail climbs and climbs. No person climbs with it, present company excepted. Steller's jays rattle noisily in the pines. My boots plunge through the snow crust time and again.

On the crest of the front range I encounter Forest Service Road 25, which runs in from the Slate Creek Divide. The track has a surface composed of equal parts of mud, compressed snow, slush and ice, courtesy of earlier runs in by Ford Broncos and the like, none of which is in evidence as I pull myself up onto the ridge. The view west from a red-rock outcrop away from the dirt road encompasses an apparent infinity of mountains, quilted with snow lying beneath the manzanita and the pines. Two ravens come in low and lift up over the exposed outcrop. As the lead bird sees me, it swings wildly over on its side for a wing stroke or two, before regaining its equanimity and original flight path. The ravens exchange a "pruk" or two as they sail out over their vast domain.

My recollection of the map of the Mazatzals tells me that I need not return the same way I came but have only to walk a short distance along the road to reach a track that parallels the one I took on the way up. I set off to find the trail, which does not appear in what my memory tells me is the correct distance. I persevere and eventually find a vandalized sign for trail 48, a number I cannot recall seeing on the map when last I examined it. Nevertheless, it seems to be headed in the appropriate direction, eastward, as required if I am ever to return to the trailhead and car.

Soon trail 48 begins to zig and zag down the mountain, with the eastern zigs becoming shorter and the western zags growing longer, until finally it is all zag. After having dropped down a good thousand feet on a slippery snow-covered trail, it is depressing to think of retracing my steps back to my ridge top starting point, but I begin to think of it anyway. Inertia keeps me pointed nervously downhill and then I see a sign for trail 45, finally a number that I do recall from my map-studying days. This trail strides out firmly to the east, paralleling Deer Creek, which is in good humor at the start, humming to itself as it sends snowmelt water plunging over gray and red rocks from pool to pool, ducking around corners and under ponderosa pines.

The afternoon drifts quietly away. The angle of descent decreases until the stream is reduced to the occasional murmur in barely moving pools separated by long stretches of dry, tumbled rocks. The snow cover becomes ever more tattered and incomplete. I put one foot in front of the other, and then repeat the formula. A flock of bushtits flits through a stand of conifers by the trail, each bird piping softly to its neighbor. The late day's sun fails to reach the bottom of the canyon, keeping Deer Creek in the shade, but far overhead three robins sail westward into the sunshine, their red breasts all aglow in a canyon where people are an exception to the rule.

The coyote in South Mountain Park

Wilderness is at least partly in the mind of the beholder, thank God, and here on the blackened rocks that protrude from a desert hillside in South Mountain Park, Phoenix, Arizona, I can manufacture a modicum of the commodity I came for. High on a mountain ridge, I can see but not hear the cars backed up on Baseline Road to the north or the caravan of traffic inching along Interstate 10 to the east.

In the park itself, a calmness pervades the afternoon. At the base of one outcrop a black-throated sparrow has built a nest in a largely leafless mallow. The nest forms a perfect circle among the undisciplined tangle of whitish dried stems. Three young sparrows flatten themselves deep in the nest cup. Their black pinfeathers have developed sufficiently to give them the appearance of birds rather than small naked mice. But their soft beaks possess a twisted yellow flange, which makes it seem as if they are grimacing as they crouch down, their mouths shut, their fear palpable.

A wistful breeze slips from one paloverde to the next on the hillside. A rock squirrel along the ridge begins a quavering trill, far different and much more prolonged than its customary piercing squeak. To the left of the squirrel, another one appears from nowhere and urgently races downhill directly to the nearest rock pile. This new squirrel ascends to the highest point on its lookout and stares intently down at the flats several hundred feet beneath the ridge.

Far below the apparently fascinated squirrel, a coyote strides confidently across the desert. With a pale coat, dark-tipped tail and sharp snout, it trots sedately between bursage and paloverde. The coyote stops, sticks its pointed muzzle under a shrub, withdraws, peers off to one side and then ambles ahead.

An orange-crowned warbler forages on the hillside above the hunting coyote. The greenish warbler travels in the company of an even drabber little flycatcher. The flycatcher sits, head cocked, while the warbler twists and turns through a paloverde, peering, peering, darting, snapping, peering, peering. The flycatcher explodes from its perch nearby to swoop out after a flying insect disturbed by the hyperactive warbler. A decisive snap of its bill announces the end of the chase.

Far below, three joggers round the bend on the dirt road several hundred yards from the coyote, and as they do, the coyote pauses to listen to them, tracking their movements and deciphering their intentions before confidently resuming its zigzagging journey across the flats. The animal continues to inspect the base of bushes here and there before dropping into a little wash that conceals it completely.

Long before the coyote vanishes, the trilling rock squirrel stops calling. Then the observer squirrel on its lookout scrambles down and bounds off between the boulders, its bushy, speckled tail undulating as the animal flows away on another errand in a life dedicated to staying alive.

It is spring now, but in the heat and haze of June and July these mountains will lose most of the comfortable greens and turn black and gray. The spindly radio towers on the highest part of the mountains will disappear amid the glare and diffusion of sunlight. On summer days and spring days alike, South Mountain always has the look of an anachronism, all odd angles and curves right next to the urban geometry of straight-lined high-rises, right-angled streets and perfectly smoothed outlying fields. South Mountain comes by irregularities honestly because it is part of the desert. But it is also a city park. The whole mountain range, 16,000 acres of rocks and paloverdes, canyon wrens and coyotes, has been converted into what Phoenicians claim is the largest municipal park in the world, one that is visited by about 1.5 million people a year.

The park owes its existence to the work of a committee of city dwellers active in the 1920s who worked with then-Representative Carl Hayden to shepherd a bill through Congress, a bill that enabled the mountain range to be transferred from federal control to the City of Phoenix in 1924 for about a dollar an acre. The foresighted city fathers then had the good sense to shut down several mining operations in South Mountain so that the land could be set aside completely for recreational purposes.

In 1924 Calvin Coolidge was president of the United States and Phoenix had a population of 30,000. One Phoenician, C.M. Holbert, became the park's first ranger when he was sworn in at age sixty-eight as a deputy sheriff in 1929. It was a relatively uncomplicated time, and Holbert operated under a superbly uncomplicated philosophy. "I made a law—no guns or axes—put it on my cards and signs and rigidly enforced it." He retired at age seventy-eight, having saved many a saguaro from the gun-toting morons of his day.

Since Holbert's era, the city's population has ballooned thirty-fold, putting greater and greater people pressure on the park, which has been protected from its users with varying degrees of effectiveness over the years. Protection failed nearly completely during one grim period of several decades, when off-road vehicles of all sizes had free rein to carve up the park. The gritty tracks they made then still criss-cross the flatter portions of South Mountain and slice straight up hillsides to mangle ridgelines. If C.M. Holbert had observed the handiwork of these vehicular vandals, he would have been utterly disheartened. Off-roaders have, however, now been declared off-limits, and police patrol the legitimate roadways on a regular basis. Few things are more gratifying in life than to see a police car, beacon flashing, pull up to call an all-terrain vehicle owner from his destructive amusements to receive a lecture and a fine.

In addition to added police patrols, the park has benefited from a volunteer group of desert admirers, the Mountaineers, who have tried to revegetate the scars left behind by off-roaders. They dig discreet terraces at right angles across ascending tracks to retain soil and gravel that inevitably erode downhill during rainstorms. And they transplant cacti and ocotillo behind the anti-erosion retention walls. But despite their hard work, the repair of the network of varicose trails in South Mountain has hardly begun.

Although off-roaders are under control for the moment, the park still comes under assorted human assaults. One recent violation takes the form of a golf course that pokes its absurdly green, wet nose into what was once the eastern end of the park. This desecration of the desert preserve was arranged by Gosnell Builders, whose huge Pointe Resort wraps around the eastern border of the park. With the connivance of the Phoenix City Council, the developers arranged a land swap that enabled them to acquire that portion of the park that they wanted in order to complete a golf course appended to their upscale resort.

A future violation waits in the wings. The state's Department of Transportation has its eye on the western end of the park where it longs to construct a multiple-laned freeway, the better to ferry people back and forth between Phoenix proper and the new bedroom community developments that have eaten up the creosote flats south of South Mountain. There tightly arrayed battalions of red-tiled roofs are interrupted with the curved shapes of baby blue swimming pools.

Despite past abuses and present encirclement, South Mountain is still big enough and rugged enough to convey a sense of the Sonoran Desert to anyone who wishes to experience it. The ridges above the back canyons rise to sufficient height to screen out much of the civilized world once you have entered this world. Hiking trails, not off-road vehicle highways, amble thinly and discreetly among the rocks. Great slabs of weathered, blackened granite shingle themselves in organized chaos down long slopes. A rock wren pops up onto one flattened boulder, bobs twice, looks right, looks left, and ducks into a crevice, sinking out of sight like a stone dropped down a mine shaft.

The trail coasts along at a moderate pitch, wandering among the scattered foothills paloverdes and then passing an odd little tree of a very different sort than the far more numerous paloverdes. Despite its modest waist-high height, the unusual tree possesses three prominent trunks that radiate outward and upward. Pale, papery bark plasters itself to the lower parts of each trunk, which are notable for their thickness at the base and the degree to which they then taper to fine twiggy points. The tapering design of the trunks apparently reminded someone of elephant trunks, thus the common name—elephant tree. However, their thickened lower tree trunks also give the plants a heavy-footed appearance, which offers yet another metaphorical justification for their name. Take your pick.

Elephant trees are not especially common in southern Arizona, and it is wonderful to have them here so close to Phoenix in South Mountain Park. Although the species also pops up on steep slopes in the Estrellas and White Tank Mountains, South Mountain's population is on the extreme northern edge of its range. Frost is a major enemy of the elephant tree. In the Estrellas and South Mountain, the tree grows almost exclusively on southeast facing slopes, the quicker to receive life-saving sunshine on winter mornings.

Heat-loving elephant trees are more numerous to the south in Sonora, Mexico, and the genus to which they are assigned (*Bursera*) contains many other species that flourish only in tropical Mexico and Central America. Whereas Arizona can claim two species of *Bursera*, one of which is so rare that it has not been relocated in recent years, there are ten species in Sonora. The various *Bursera* trees are noteworthy not just for their odd shapes, but also because of their intensely aromatic character. Each little elephant tree on a South Mountain slope surrounds itself with a delicious scent emanating from its leaves and bark. Yellow resin bubbles through small openings on the surface of limbs and trunks, sometimes forming thin coils that dangle from the tree.

The aroma of an elephant tree has been described as similar to that of turpentine, but this is much too unkind. The odor is certainly resinous, highly pervasive, a bit astringent, but it is fundamentally pleasant, perhaps because it seems softened with a hint of lemon. Ancient and modern Mexicans must have agreed that *Bursera* resin produced a pleasing, rather than repellent, odor because then and now the resin of these trees has been collected to produce *copal*, an incense used in a wide range of occasions, some religious in nature. Pop a chunk of elephant tree resin in a campfire and you will see why.

Products of the elephant tree also played many roles for people of an entirely different culture, the Seri Indians of the coastal Sonoran Desert of northwestern Mexico. Among other things, the Seri produced a tea of elephant tree leaves or twigs. The tea was used on vision quests conducted during three or four days of fasting in the wilderness in which the quester sought spiritual enlightenment.

The Seri invented uses for all the parts of elephant trees, uses concerning everything from the supernatural world to the most mundane aspects of life. The tree contributed materials for boat caulking, face painting, and fire making as well as for shampoo, headbands and medicines. The Seri killed head lice with a solution containing crushed elephant tea fruit, and they treated gonorrhea with a tea made of boiled elephant tree bark. Men who fished standing in the sea fashioned a belt of the twigs to repel sharks, which perhaps found the scent of the resins unpleasant, or so the Seri hoped.

Bursera trees invest heavily in the production of resins for a practical reason of their own, the defense of their tissues against plant consumers. Just as turpentine is a natural plant product used to poison enemies of certain pines, so, too, the terpenes and other aromatic substances contained in the tissues of Bursera trees act as toxins against would-be herbivores. Consider the Central American species Bursera schlechtendalii, whose very name suggests that it is not to be taken lightly. This relative of the elephant tree anticipated the invention of pesticide sprayers; pluck an entire leaf and a fine spray of terpenes blasts out from the remaining bit of petiole for three to four seconds. The squirt of resinous aerosol travels up to six inches and probably deters browsers from nibbling more leaves from one of these trees. Terpenes may be pleasantly aromatic, but they do not taste good to most animals for the simple reason that they are generally poisonous. Turpentine, to pick an example, is not a recommended digestive.

In addition to the "squirt-gun response," B. schlechtendalii has another pesticide trick that probably works against smaller leaf eaters. When a small-mouthed consumer, such as the larva of leaf-eating beetles, bites into a leaf, slicing a section from it, the leaf responds by releasing a surge of terpene-filled fluid from the wound edge. This "rapid bath" response pours the fluid onto the body of the caterpillar, much to its discomfort.

Both responses rely on a system of resin canals that reticulate through the leaves and stems of the tree. Sufficient pressure exists on the fluids in the canals to permit the explosive or rapid release of defensive compounds when herbivores bite into stems and leaves.

The elephant tree's protective plumbing is less elaborate. Snip an entire pinnate leaf of Bursera microphylla and nothing especially dramatic happens, although a droplet of highly aromatic resin does well up at the wound. If touched, the fluid quickly wets adjacent leaves or fingers and would probably spread equally rapidly over the body of a leaf-eating caterpillar of some sort. I guess, and it is a guess only, that the elephant tree's moderate response to leaf injury suffices to discourage most potential herbivores. The leaves on the South Mountain elephant trees seem remarkably free from insect damage of any sort. Each leaf's set of flat, neatly arrayed leaflets are largely unblemished, so that overall the foliage looks intact and healthy in a feathery way. One good freeze, however, can kill the lot, curling the leaflets and removing much of the plant's perfume.

The fruits of the elephant tree hang close to the limb and twigs from small curved petioles. They look like miniature plums with a waxy bloom on their red-purple surfaces. Doves find the fruits highly appealing, eating quantities to secure the thin rind and discarding the large orange seed within. From these dispersed seeds, new elephant trees occasionally arise.

Judging from the enthusiasm of doves for the fruits, this part of the tree is not toxic. I suffered no ill-effects when tasting one. The Seri, too, consumed the fruit without injury, chewing the fruit to induce the flow of saliva and thereby help alleviate the effects of thirst when they were far from water. The children of the Seri also used the fruit as substitute peas for peashooters made of a locally available hollow reed. The fruits possess just the right size and firmness to be propelled with authority from a peashooter.

I had lived in Arizona for nearly two decades before I first noticed elephant trees, coming across a specimen or three during a ramble through South Mountain, where 274 different plant species have been recorded. At that time I recognized the tree as being strangely different from the ones with which I had grown familiar, but I did not get around to putting a name on the plant for several more years. Still more time passed until Joe McAuliffe pointed me toward the book, *People of the Desert and Sea*, which catalogs the spectrum of uses that were apparent to Seri in the fruit, leaves, twigs and bark of the elephant tree. Unlike the Seri, I derive no practical value from the presence of elephant trees in my part of the Sonoran Desert. Nevertheless, I am pleased that it is here and glad to have finally made its acquaintance. For this I thank the handful of people who acted nearly seventy years ago to preserve South Mountain, providing a protected desert island in an increasingly urban world, a place where the natural incense of elephant trees still hangs in the air, enriching the senses and providing an aromatic memorial to Seri life as it used to be.

December rain

A fog and light drizzle marks the end of another year in the Sonoran Desert. Today's weather has its roots in a disturbance that has scuffled up from Baja California, passing over the traditional homeland of the Seri and thousands of elephant trees. The moderate temperatures and soft rain are a Christmas gift from Mexico. The saguaros on Usery Mountain come looming out of the grayness, at first pale and blurred, and then clearer and better defined as I approach them. In the still, cool air of midmorning, the fine mist hangs in curtains that droop lower and lower as they fall ever so slowly to the earth. Droplets of water ornament every paloverde twig and every bursage leaf. Paloverde #17 has rarely looked as alive as it does today.

The rains over the past few weeks have restored color to many desert plants. The once brown, twisted resurrection plants are now truly resurrected with unfolded leaves, richly green; the blackened bits of moss that cling to north-facing rocks and slopes have also been transformed into smooth mounds of green velvet, round and plump.

Even plants that have not made a complete metamorphosis in the winter rains are changed for the better. The flattened, curved spines of the fishhook barrel cacti, which are normally a pale, subdued red, are now just this side of crimson. The swollen, cylindrical cacti push up against black rock faces covered with lichens. Flattened, but slightly corrugated, the lichens press tight to their boulders, like rock climbers on an ascent. The lichens also seem rejuvenated by the day, offering a brighter-than-usual palette of lively colors, chartreuse, green-white, blue-green, orange-umber.

Some things, however, have not survived the hard freeze of the previous week, and they remain blackened, uncolored by the rain. Many brittlebushes, especially the smallest ones, froze, then thawed, and now are left with dead curled leaves. No sign at all exists of the armies of seedling shrubs that sprouted after the late summer rains, flourished for a time, and then went down one by one during a dry spell until the bitter cold provided a final coup de grace.

Peccaries have been out walking in the rain in search of living edible plants. I come across four or five widely scattered fresh droppings. Peccary footprints, deeply impressed in the soft, moist soil, track the hillsides and the ridgelines. One trail of cloven prints cuts straight across a sandy wash and marches up the bank past a trio of teddy-bear chollas.

Teddy-bear chollas usually seem to have their own internal light source, which generates a creamy radiance capable of competing with the desert sun. Today the sun has taken a rain check and the chollas have lost all hint of yellow or cream. Instead, they emit a pure winter white, heightened by the combined reflections from the droplets hung on one thousand spines.

A wind-driven heavier drizzle briefly tattoos my hat and all that surrounds me, including a dead paloverde whose rain-soaked wood has blackened, revealing its cracked and fractured form. A tomatillo drapes its thin branches over the dead body of the tree.

Mist has pooled in the great valley that separates the Userys from Red Mountain. As the day progresses, the cottony white rises and falls whimsically; the wind picks up streamers and carries them over the ridges before abandoning them to drift aimlessly across the mountainsides. The mist descends and thickens, concealing all, and then the air clears again; the sun nearly penetrates the cloud mass for a minute or two, enough to cast an insecure light over a broad patch of the desert, only to sink back into obscurity.

As a fine drizzle saturates the air again, the saguaros collect the moisture, which slips easily down their smooth skin until interrupted by some asymmetry. There the water gathers itself together in fat drops before leaping free to fall upon the already soaked ground.

Orange streaks and patches intermingle with green on the swollen trunks of the larger saguaros. A yellow mold clings to the grooved wood of a long-deceased but still standing ironwood. In the muffled silence of the day, a rock wren flutters without a sound from beneath one desert shrub to disappear under another. The bursage that it left behind shelters a sedentary young foothills palo-verde, whose thin green branches have just begun to emerge from within the concealing canopy of its unwitting protector.

A small, white jaw bone juts out of unsheltered ground. The sharp slicing back teeth and miniature but feisty canines say "carnivore," but which one? A complete lower mandible three inches long, it is of a size appropriate for a baby skunk or perhaps a ringtail cat that not so many months ago poked among the rock piles on the mountain.

The big wash in the hidden valley on the southeastern flank of the Userys has been restructured by the rains. A thin layer of pale gravel and sand has obliterated the few signs of human traffic accumulated over the previous year. The fresh coat of sand conceals my old footprints and those of a couple of others who reached this isolated wash. On the surface of the rain-soaked sand there are no cigarette butts, no far-blown fragments of newsprint, no broken bits of balloon noosed to thin white strings. The desert seems to be trying once more to regain its equilibrium, to start the new year fresh, to put people out of its collective mind. Perhaps someday we will help it succeed.

References

A natural history

Betancourt, J.L., T.R. van Devender, and P.S. Martin (eds.). 1990. *Packrat Middens*. Tucson, University of Arizona Press.

Turner, R.M., and J.E. Bowers. 1988. Long-term changes in populations of *Carnegiea gigantea*, exotic plant species, and *Cercidium floridum* at the Desert Laboratory, Tumamoc Hill, Tucson, Arizona. In *Arid Lands: Today and Tomorrow: Proceedings of an International Conference*, Whitehead, E.E., C.F. Hutchinson, B.N. Timmerman, and R.G. Varady (eds.). Boulder, Westview Press.

Army ants

Topoff, H.R. 1984. Social organization of raiding and emigrations in army ants. *Advances in the Study of Behavior* 14:81–126.

The birth of a paloverde

McAuliffe, J.R. 1986. Herbivore-limited establishment of a Sonoran desert tree: *Cercidium microphyllum*. *Ecology* 67:276–80.

McAuliffe, J.R. 1990. Paloverdes, pocket mice, and bruchid beetles: interrelationships of seeds, dispersers, and seed predators. *Southwestern Naturalist* 35:329–37.

Shreve, F. 1917. The establishment of desert perennials. *Journal of Ecology* 5:210–16.

The miner's cat

Ary, T.S. et al. (National Legal Center for the Public Interest). 1989. The mining law of 1872: A legal and historical analysis. NLCPI, Washington DC.

Blair, R. 1975. *Tales of the Superstitions.* Tempe, Arizona Historical Foundation.

Where did all the glyptodonts go?

Anderson, A. 1989. *Prodigious Birds.* Cambridge University Press, New York.

Diamond, J. 1990. Bob Dylan and moas' ghosts. *Natural History* (Oct.):26–32.

Grayson, D.K. 1977. Pleistocene avifaunas and the overkill hypothesis. *Science* 195:691–93.

Martin, P.S. 1990. 40,000 years of extinctions on the "planet of doom." *Palaeogeography, Palaeoclimatology, Palaeoecology (Global and Planetary Change Section)* 82:187–201.

Steadman, D.W. and P.S. Martin. 1984. Extinction of birds in the late Pleistocene of North America. In *Quaternary Extinctions: A Prehistoric Revolution,* Martin, P.S. and R.G. Klein (eds.). Tucson, University of Arizona Press.

Sutcliffe, A.J. 1985. *On the Track of Ice Age Mammals.* London, British Museum (Natural History).

Thirty-eight Apaches

Cozzens, S.W. 1891. *The Marvellous Country: or, Three Years in Arizona and New Mexico, the Apaches' Home.* Boston, Lee and Shepard.

Debo, A. 1976. *Geronimo*. Norman, University of Oklahoma Press.

Ferg, A. (ed.) 1987. *Western Apache Material Culture*. Tucson, University of Arizona Press.

Schurr, T.G. et al. 1990. Amerindian mitochondrial DNAs have rare Asian mutations at high frequencies, suggesting they derived from four primary maternal lineages. *American Journal of Human Genetics* 46:613–23.

Sweeney, E.R. 1991. *Cochise*. Norman, University of Oklahoma Press.

Thrapp, D.L. 1967. *The Conquest of Apacheria*. Norman, University of Oklahoma Press.

Williams, R.C. et al. 1985. GM allotypes in Native Americans: evidence for three distinct migrations across the Bering Land Bridge. *American Journal of Physical Anthropology* 66:1–19.

The last Indian war?

Lyman, A.R. 1962. *Indians and Outlaws: Settling of the San Juan Frontier*. Salt Lake City, Bookcraft.

McPherson, R.S. 1985. Paiute Posey and the last white uprising. *Utah Historical Quarterly* 53:248–67.

Confessions of a cactus-hugger

Seeber, L.C. 1971. *George Elbert Burr, 1859–1939*. Flagstaff, Northland Press.

Abert's towhees and other opportunists

Brandt, H. 1951. *Arizona and Its Bird Life*. Cleveland, Bird Research Foundation.

Diamond, J.M. 1985. Rapid evolution of urban birds. *Nature* 324:107–8.

Emlen, J.T. 1974. An urban bird community in Tucson, Arizona:

derivation, structure and regulation. *Condor* 76:184–97.

Phillips, A.R., J. Marshall, and G. Monson. 1964. *The Birds of Arizona.* Tucson, University of Arizona Press.

Rosenberg, K.V., S.B. Terrill, and G.H. Rosenberg. 1987. Value of suburban habitats to desert riparian birds. *Wilson Bulletin* 99:642–54.

Playing God with the white-winged dove

Arnold, L.W. 1943. A study of the factors influencing the management of and a suggested management plan for the western whitewinged dove in Arizona. Phoenix, Arizona Game & Fish Commission.

Brown, D.E. 1989. *Arizona Game Birds.* Tucson, University of Arizona Press.

Cattle free in 1893

Bahre, C.J. 1991. *A Legacy of Change.* Tucson, University of Arizona Press.

Hastings, J.R. and R.M. Turner. 1965. *The Changing Mile: An Ecological Study of Vegetation Change with Time in the Lower Mile of an Arid and Semiarid Region.* Tucson, University of Arizona Press.

Moseley, J.C., E.L. Smith, and P.R. Ogden. 1990. *Seven Popular Myths about Livestock Grazing on Public Lands.* Moscow, University of Idaho, Forest, Wildlife and Range Experiment Station.

Turner, R.M. 1990. Long-term vegetation change at a fully protected Sonoran Desert site. *Ecology* 71:464–77.

Mountain lion mathematics

Brown, D.E. 1985. *The Grizzly in the Southwest: Documentary of an Extinction.* Norman, University of Oklahoma Press.

Shaw, Harley. 1989. *Soul Among Lions.* Boulder, Johnson Books.

Cowpies

Bostick, V. 1990. The desert tortoise in relation to cattle grazing. *Rangelands* 12: 149–51.

Jarchow, J.L. 1984. Veterinary management of the desert tortoise *Gopherus agassizii* at the Arizona-Sonoran Desert Museum: A rational approach to diet. *Proceedings of the Desert Tortoise Council* 1984:83–94.

Jones, S.C. 1990. Colony size of the desert subterranean termite *Heterotermes aureus* (Isoptera: Rhinotermitidae). *Southwestern Naturalist* 35:285–91.

Whitford, W.G., Y. Steinberger, and G. Ettershank. 1982. Contributions of subterranean termites to the "economy" of Chihuahuan desert ecosystems. *Oecologia* 55:298–302.

Peccaries

Sowls, L. 1984. *The Peccaries*. Tucson, University of Arizona Press.

Death in a saguaro forest

Abou-Haidar, F. 1989. Influence of livestock grazing on saguaro seedling establishment. Master's thesis, Arizona State University.

Hastings, J.R. and R.M. Turner. 1965. *The Changing Mile: An Ecological Study of Vegetation Change with Time in the Lower Mile of an Arid and Semiarid Region*. Tucson, University of Arizona Press.

Randolph Canyon and Burro Creek

Marston, E. 1991. Rocks and hard places. *Wilderness* 54:38–45.

Minckley, W.L. and T.O. Clark. 1984. Formation and destruction of a Gila River mesquite bosque community. *Desert Plants* 6:23–30.

The masked bobwhite rides again

Brown, D.E. and D.H. Ellis. 1984. Masked bobwhite recovery plan. Albuquerque, U.S. Fish and Wildlife Service.

Brown, H. 1900. Conditions governing bird life in Arizona. Auk 17:31–34.

Brown, H. 1904. Masked bobwhite (Colinus ridgwayi). Auk 21: 209–13.

The coyote in South Mountain Park

Becerra, J.X. and D.L. Venable. 1991. Rapid-terpene-bath and "squirt-gun" defense in Bursera schlechtendalii and the counterploy of chrysomelid beetles. Biotropica 22:320–23.

Daniel, T.F. and M.L. Butterwick. 1992. Flora of the South Mountains of south-central Arizona. Desert Plants 10:99–119.

Felger, R.S. and M.B. Moser. 1985. People of the Desert and Sea. Tucson, University of Arizona Press.

Johnson, M.B. 1992. The genus Bursera (Burseraceae) in Sonora, Mexico and Arizona, U.S.A. Desert Plants 10:126–43.

Acknowledgments

I am grateful to the community of desert biologists upon whose research and writings I have drawn heavily in producing my book. I have benefited from conversations with many of my colleagues in the Department of Zoology at Arizona State University. The department is blessed with a substantial number of opinionated, beer-drinking field ecologists who graciously associate with me, among them Jim Collins, Stuart Fisher, Mink Minckley, Mike Moore, Dave Pearson, Ron Rutowski and Glen Walsberg. Thanks also go to Conrad J. Bahre, David E. Brown, Paul S. Martin, Joseph R. McAuliffe and Raymond M. Turner for their contributions to understanding what the West is all about and how it got that way. Finally, my wife Sue and sons Joe and Nick have helped make my life in the West full of changes and interest.

Index

Abert's towhees, 82–84, 90, 93
Alcorn, Stanley, 138–39
Anasazi culture, 61–64
Animal Damage Control
    agency, 125, 128
Antiherbivore response, 170–71
Ants: army, 10–13; harvester,
    11–13
Apache Indians, 51, 53–61, 150.
    See also Indian reservations
Apache Junction, 79–81
Audubon Christmas counts, 16

Backpacking, 36–37
Bahre, Conrad, 113–14
Ballantine Canyon, 160–62
Bascom, Lt. George N., 56–57
Beetles: burying, 14–15; dung,
    108; seed, 26
Behavioral flexibility, 86–88
Beston, Henry, vii, ix
Bison, 50
Black bears, 122–24, 161–62

Black-tailed gnatcatcher, 13–16,
    21, 84, 86
Bostick, Vernon, 131–33
Brandt, Herbert, 82–83
Brown, Herbert, 147–48, 150
Bureau of Land Management,
    U.S., 61–63, 98, 106–7, 115–16,
    144, 146–147
Burr, George Elbert, 79–81
Burro Creek, 144, 146–47

Canyon wren, 134, 136, 143
Cattle grazing: economic im-
    portance of, 98–99, 128; and
    native animals, 130–32, 135–
    36, 149; and native plants,
    100–2, 108–17, 139–41
Cattle populations, in South-
    west, 109–11
Cave Creek, 97, 100, 116–21
Chief Posey, 64–69
Chiricahua Mountains, 58, 61,
    71, 97–98, 103, 107

Climate change, 6–7, 44–45
Clovis culture, 46–47, 49–50, 53
Cochise, 56–57
Colonization: of Americas,
    45–46, 53; of Southwest, 42,
    54–56
Conservation projects, 136,
    151–53, 166–67
Coronado National Forest,
    118–21
Cow dung, 97, 129–33
Coyotes, 10, 125, 165

DeConcini, Dennis, 146
Desertification, 109–13
Desert streams, 144
Desert tortoise, 131–33
Diamond, Jared, 48, 86

Earth First!, 105
Ecological change, 6–8, 98,
    109–13, 136–37
Eco-terrorism, 105, 124
Elegant trogon, 116–17
Elephant trees, 168–72
El Niño, 17
Emlen, John T., 84–85
Environmental impact state-
    ments, 117–21
Extinction, 44–50, 93, 148

Farming, 53, 77, 91
Feeding behavior: carrion eat-
    ing, 14–15; of desert tortoise,

131–33; seed eating, 25–27, 85;
    of termites, 130–31
Fire suppression, 114
Fish and Wildlife Service, U.S.,
    151–52
Floods, 92, 144–47
Forest Service, U.S., 106–7,
    115–21, 123, 144–45, 162–63
Free, Mickey, 57, 59

Genetics, of American Indians,
    53
Geronimo, 52, 54, 59–61, 64
Gila monster, 157
Grasslands, Arizona, 109, 136–
    37, 149–54
Grayson, Don, 48–49
Grazing leases, vii, 34, 97–99,
    103, 106, 120, 124–25, 139, 146,
    162
Grazing subsidy, 106, 119
Ground sloths, 43

Hastings, James, 110, 112, 138, 159
Hohokam culture, 42, 77
Horses, 54
Hunting: of doves, 90–91; of
    megafauna, 44–50

Indian reservations: Chiricahua
    Apache, 58; Fort McDowell,
    42, 51; San Carlos Apache, 51;
    White Mesa Ute, 67
Introduced species, 85–88, 92

Jackrabbits, 27–28
Javelinas. *See* Peccaries
Jeffords, Tom, 58
Jorgenson, Clive, 6

Kangaroo rats, 69–73

Lackner, Eddie, 122–25
Linguistics, 53
Littering, 34–35, 41
Lockard, Bob, 69–73
Lost Dutchman Mine, 39
Lowe, Charles, 138
Lyman, Albert R., 68

McAuliffe, Joseph, 24–25, 27
McCormack and Company, 35
McPherson, Robert, 67
Mammoths, 43
Mangas Coloradas, 55–58
Maori culture, 48
Marshall, Joe, 87
Martin, Paul, 43–50
Masked bobwhite, 147–54
Mazatzal Mountains, 162–64
Megafauna, 44–50
Mesquites, 82, 90, 107–8, 136,
    146–47
Mexican immigrants, 73, 75
Mexican settlers, 54–55
Mining, 33, 39–41
Mining Law of 1872, 39–40
Moas, 47–48
Monson, Gale, 87

Mormons, 64, 66–69
Mortality, bird, 14–15
Mountain lions, 121–28

Na-Dene, 52–53
Navajo, 52–53
New Zealand, 47–49
Nutting, William L., 131

Off-road vehicles, 3, 166–67
Oman, Donald, 144–46
Overgrazing, 109–11, 139, 149
Overkill, 47–50

Paiutes, 64–68
Paloverde, 7; age, 21; seed ger-
    mination, 25, 27; seedling
    mortality, 27
Panthers, 126
Passenger pigeons, 93
Peccaries, 17, 134–36
Peloncillo Mountains, 59, 71
Phillips, Allen, 87
Phoenix, population of, 79
Pinacate Mountains, 111–13
Pioneers, 5–6, 53, 67
Pocket mice, 24–26
Population growth, human,
    76–79, 127
Predator avoidance, 70
Predator control, 121–28
Prickly pear cactus, 136
Puebloan culture, 53

Randolph Canyon, 101, 141–43
Range management, 101–2,
    114–16, 131, 141
Recreation, 36–37, 120
Ringtailed cat, 37–39
Riparian habitats, 82–84, 88,
    146–47
Rock squirrel, 165
Rosenberg, Ken, 83–84
Roth, Vince, 103–5

Saguaro cactus, 9–10, 174;
    frost damage, 4–5; fruits, 89;
    growth, 139–41; mortality,
    20–21, 137–41, 157–60; nest
    sites in, 87
Saguaro National Monument,
    138–41
Salt cedar, 92
Schilling's Company, 34–35
Seed caching, 24–27
Seri Indians, 169–71
Shreve, Forrest, 21, 24
Sonoran desert: age of, 5–6;
    seasonal change in, 17
South Mountain, 164–72
Southwestern Research Station,
    103–4
Starlings, 86–88
Steadman, David, 49
Stream downcutting, 109
Superstition Mountains, 36, 39,
    80–81, 117, 128, 133–35

Superstition Wilderness Area,
    40–41, 79, 81, 101, 142–43
Symms, Stephen, 145
Synar, Mike, 106–7

Tarantula hawk wasps, 22
Target shooting, 11
Teddy-bear cholla, 8–9, 160, 173
Termites, 130–33
Turner, Raymond, 110, 112, 138,
    141, 159

U.S. Bureau of Land Man-
    agement, 61–63, 98, 106–7,
    115–16, 144, 146–147
U.S. Fish and Wildlife Service,
    151–52
U.S. Forest Service, 106–7, 115–
    21, 123, 144–45, 162–63
Urban birds, 83–88
Usery Mountains, 3, 7, 17, 29,
    33–35, 78, 117, 157–60, 172–74

Wells, Phil, 6
White-winged dove, 89–93
Whitford, Walter, 130
Woodrat middens, 5–6

Yavapai Apaches, 42, 51

Zwinger, Ann, 19

About the author

JOHN ALCOCK is Regents' Professor of Zoology at Arizona State University and the author of *Animal Behavior: An Evolutionary Approach*, the most widely used textbook in animal behavior. For much of his career he has studied insect behavior in the Sonoran Desert of central Arizona, where he has developed a great admiration for the desert and its natural inhabitants. Alcock has conveyed his enthusiasm for desert biology to the reading public in magazine articles for *Natural History* and *Arizona Highways* and in the books *Sonoran Desert Spring*, *The Kookaburras' Song*, and *Sonoran Desert Summer*.